The Complete Guide to
DIGESTIVE HEALTH

Josephine Spire

Emerald Guides
www.straightforwardco.co.uk

Emerald Guides

© Josephine Spire 2022

Josephine Spire has asserted the moral right to be identified
as the author of this work.

ISBN 978-1-80236-060-8

Printed by 4edge www.4edge.co.uk
Cover design by BW Studio Derby
Typeset by Frabjous Books

Whilst every effort has been made to ensure that the information
contained within this book is correct at the time of going to press,
the author and publisher can take no responsibility for the errors
or omissions contained within.

CONTENTS

PART 3 – THE IMPORTANCE OF NUTRITION IN DIGESTION

PART 4 – DISEASES OF THE DIGESTIVE SYSTEM

INTRODUCTION

The digestive system is an especially important part of the body and without it, it would be impossible to get the nutrients we need to fuel our bodies for energy, growth, and repair.

Digestion is a mind-blowing and magnificent process which we tend to take for granted until it no longer works properly. To make sure that your body's digestive system has your support, you must eat healthily to avoid digestive problems that can occur due to a poor diet, unhealthy habits, and sensitivity to certain foods. It is often stated that eating is one of life's greatest pleasure, however, very few people know what exactly happens to that food once they have ingested it!

The fact that we can heal our bodies and indeed our digestive systems through eating the right diet is the best kept secret by the medical world. Truth is, we are what we eat. Our physical and mental well-being is linked to what we eat and drink. The nutritional content of what we eat in our diet determines the composition of our blood, tissue, organs, cell membranes, hormones, bone marrow, hair and skin. To fully understand how your body works, it is important that you acquire a basic knowledge of how your digestive system functions. A healthy diet not only aids better health for your digestive tract, but it also keeps the doctor at bay and makes you feel great.

Most people do not want to talk about their digestive issues, and rarely seek doctor's advice for the minor harmless conditions like heartburn, constipation, diarrhea, bloating and so forth. However, some digestive symptoms like diarrhea if left untreated for lengthy periods, can lead to more serious chronic diseases. And while some digestive problems can be managed with over-the-counter medicines and lifestyle changes, others can be profoundly serious and life threatening, requiring you to seek medical help from your doctor. This book is a perfect balance of science and practical advice. It explains clearly how your digestive system works and what happens when it does not function properly-providing everything you need to take control of your overall health through close, careful attention to your digestive system.

* * *

PART 1

The Digestive Process

The digestive system is responsible for taking in and breaking down foods, and turning them into nutrients for energy. These nutrients are essential for the body to function, grow and repair. The food and drink we consume has to be broken down into smaller nutrients which are made up of carbohydrates, proteins, fats, and vitamins, before the blood can absorb them to be utilized by the body. The main processes of the digestive system are ingestion of food, secretion of fluids and digestive enzymes, mixing and movement of food and waste, digestion, or breakdown of food into smaller pieces, absorption of nutrients and excretion or elimination of waste by-products.

Ingestion: Obtaining nutrition and energy from food is a multi-step process. The first step is ingestion, the act of taking in food and drink through the mouth.

Secretion: The digestive system secretes around 7 liters of fluids every day. In the gastrointestinal tract, these fluids include saliva, mucus, hydrochloric acid, bile, and enzymes. Saliva from the salivary glands in the mouth moistens dry food and contains salivary amylase, a digestive enzyme that begins the digestion of carbohydrates. Mucus aids as a protective barrier and lubricant inside of the GI tract. Hydrochloric acid performs two functions in the digestive tract: the first function is to digest food chemically and the second is to protect the body by killing bacteria which may be present in our food. Bile on the other hand is used to emulsify lipids into tiny globules for easy digestion.

Churning and Movement: Mixing and moving the ingested food involves three processes which include swallowing, peristalsis, and segmentation. The swallowing process utilizes both smooth and skeletal muscles in the mouth, tongue, and pharynx to push food into the pharynx, and into the esophagus. Peristalsis involves transporting food and other substances down the GI tract via muscular waves that travels the length of the GI tract. It takes numerous peristaltic waves for food to travel from the esophagus to the stomach and intestines and reach the other end of the GI tract.

Segmentation is another process that occurs only in the small intestine. They occur as short segments of the intestine contract. Segmentation helps to increase the absorption of nutrients and other materials by mixing food, thereby increasing its contact with the walls of the intestine.

Digestion: Digestion is the process of turning large particles of food into smaller particles. This process involves both mechanical and chemical processes. Mechanical digestion is the physical breakdown of large particles of food into smaller pieces. This begins with the chewing of food by the teeth and continues through the muscular mixing of food by the stomach and intestines. Bile produced by the liver and stored in the gallbladder is also used to chemically break fats or lipids into smaller globules.

Mechanical digestion is accompanied by chemical digestion. Food is chemically digested as larger and more complex molecules are being broken down into smaller molecules that are much easier to absorb. Chemical digestion also begins in the mouth with the salivary glands secreting saliva that contains salivary amylase. The amylase helps to split complex carbohydrates into simple carbohydrates, while the acid and enzymes found in the stomach continue with the chemical digestion.

However, the bulk of chemical digestion takes place in the small intestine aided by the pancreatic secretions. This pancreatic juice contains numerous enzymes which are proficient in digesting lipids, carbohydrates, proteins, and nucleic acids. By the time the food leaves the first section of the small intestines in the duodenum part, it has been reduced to its basic chemical building blocks; composed of fatty acids, amino acids, monosaccharides, and nucleotides.

Absorption: Once food has been broken down into small particles, it's ready to be absorbed. Absorption begins in the stomach with molecules like water, which is then absorbed directly into the bloodstream. Most of the absorption takes place in the walls of the small intestine. The folds in the intestine allows for greater surface area, facilitating the food to stay in longer contact with the intestinal wall. The small blood and lymphatic vessels located in the intestinal wall pick up the molecules and carry them the rest of the body. The large intestine is also involved in the absorption of water, vitamins B and K before the feces leave the body through elimination.

Excretion: The final function of the digestive system is the excretion or elimination of waste in a process known as defecation. Defecation removes indigestible substances from the body so that they do not accumulate inside the lower gastrointestinal tract. This process is controlled voluntarily by the brain, and must occur on a regular basis to prevent the backup of indigestible materials.

How digestion works

Digestion involves the mixing of food, its movement through the digestive tract, and the chemical breakdown of the large molecules of food into smaller molecules. Digestion begins in the mouth when we chew and swallow, and is completed in the small intestine. The chemical process varies for various kinds of food. For instance, fat takes longer to digest than other foods.

Production of digestive juices

The glands that act first are in the mouth and are called the salivary glands. Saliva produced by these glands contains an enzyme that begins to digest the starch from food into smaller molecules. The next set of digestive glands is in the stomach lining. They produce stomach acid and an enzyme that digests protein. In most people, the stomach mucosa is able to resist the juice, although food and other tissues of the body cannot.

After the stomach empties the food and juice mixture into the small intestine, the juices of two other digestive organs mix with the food to continue the process of digestion.

One of these organs is the pancreas. It produces a juice that contains a wide array of enzymes to break down the carbohydrate, fat, and protein in food. Other enzymes that are active in the process come from glands in the wall of the intestine or even a part of that wall.

The liver produces yet another digestive juice – bile. The bile is stored between meals in the gallbladder. At mealtime, it is squeezed out of the gallbladder into the bile ducts to reach the intestine and mix with the fat in our food. The bile acids dissolve the fat into the watery contents of the intestine, much like detergents that dissolve grease from a frying pan. After the fat is dissolved, it is digested by enzymes from the pancreas and the lining of the intestine.

Each part of the digestive system has a key role to play in the digestion of food. The process of digestion starts when food enters the mouth where it is chewed into tiny pieces lubricated by saliva and moved about by the tongue. When the food becomes soft, it is swallowed through the throat which is also known as the pharynx and passes into the esophagus. From the esophagus, the food travels to the stomach where is it stored, churned, and mixed with gastric juices secreted by its lining. Digestion continues on to the small and large intestines. From there it goes to the rectum and ends in the anus for removal. It is now well known that the digestive tract hormones play a key role in controlling food intake and energy expenditure. The gut is the body's largest hormone-producing organ, releasing more than twenty different peptide hormones, some of which target the brain to regulate appetite and influence the pleasure of eating.

Why is digestion important?

When we eat food, it is not in a form that the body can use straight away. Our food and drink must be changed into smaller molecules of nutrients before they can be absorbed into the blood stream and carried to cells throughout the body. Digestion is the process by which food and drink are broken down into their smallest parts so that the body can use them to build and nourish cells and to provide energy. As a result, through this process, the body absorbs nutrients while the digestive system gets rid of the waste.

Digestion is important because the body needs nutrients from food in our diet to work efficiently and stay healthy. These nutrients are: proteins, carbohydrates, fats, vitamins, minerals and water. Your digestive system breaks down these nutrients into small parts for the body to absorb and use for growth, energy, and cell repair.

Our digestive system is not just responsible for converting food into nutrients that our bodies need, but it is also in charge of expelling what we do not need. If our body is unable to extract any nutrients, then what is left is treated as waste and flushed out of ourbodies as either urine or feces. If we suffer from poor digestion, then this process can be slowed down, causing constipation or food can be passed too quickly through our digestive tract causing diarrhea and dehydration. The build-up of waste products in our system can also prompt the release of toxins into our system and lead to a surge in the population of unfriendly bacteria in our stomach, stimulating symptoms such as bloating, irritation, and abdominal pain.

How is the digestive process controlled?
Nerve regulators
The hormones and nerves in your body work together to help control the digestive process. Signals flow within and through the GI tract to the brain. The digestive system has a complex system of food movement and secretion regulation, which are vital for its proper function. Movement and secretion are regulated by long reflexes from the central nervous system (CNS), short reflexes from the enteric nervous system (ENS), and reflexes from the gastrointestinal system (GI) peptides that work in harmony with each other. In addition, there are three overarching reflexes that control the movement, digestion, and removal of food and food waste:

- The enterogastric reflex
- The gastrocolic reflex
- The gastroileal reflex

Long reflexes to the digestive system involve a sensory neuron that sends information to the brain. This sensory information can come from within the digestive system, or from outside the body in the form of emotional response, danger, or a reaction to food.

These alternative sensory responses from outside the digestive system are also known as feedforward reflexes. Emotional responses can also trigger GI responses, such as the butterflies in the stomach feeling when nervous. Short reflexes, on the other hand, are orchestrated by intrinsic nerve plexuses within the alimentary canal wall. These two plexuses and their connections were introduced earlier as the enteric nervous system.

Short reflexes regulate activities in one area of the digestive tract and may coordinate local peristaltic movements and stimulate digestive secretions. For example, the sight, smell, and taste of food initiate long reflexes that begin with a sensory neuron delivering a signal to the medulla oblongata. The response to the signal is to stimulate cells in the stomach to begin secreting digestive juices in preparation for incoming food. In contrast, food that distends the stomach initiates short reflexes that cause cells in the stomach wall to increase their secretion of digestive juices.

Control of the digestive system is also maintained by the enteric nervous system (ENS), which can be thought of as a digestive brain that helps to regulate motility, secretion, and growth. The enteric nervous system can function as a fast, internal response to digestive stimuli. When this occurs, it is called a short reflex.

Hormone regulators

These hormones work in association with the GI tract's extensive nervous system (Enteric nervous system) and play a coordinating role in the control of appetite, the digestion of food, the regulation of energy balance and the maintenance of blood glucose levels. The gut continuously sends information to the brain regarding the quality and quantity of the food that is consumed.

The role that some of these hormones play includes:

- Ghrelin is produced in the stomach, and its function is to tell the brain that the body must be fed. It increases appetite.
- Gastrin is produced in the stomach when it is stretched and stimulates the release of gastric juice rich in pepsin and hydrochloric acid.
- Secretin is produced in the duodenum and has the effect of stimulating the pancreas to produce alkaline secretions as well as slowing the emptying of the stomach.
- Cholecystokinin (CCK) is produced in the duodenum. It reduces appetite, slows down the emptying of the stomach and stimulates the release of bile from the gall bladder.
- Peptide YY (PYY) is produced in the last part of the small intestine known as the ileum as well as parts of the large intestine. It plays a role in slowing down the passage of food along the gut, which increases the efficiency of digestion and nutrient absorption after a meal.
- Glucagon-like peptide 1 (GLP-1) is produced in the small intestine and colon and has multiple actions including inhibition of gastric emptying and appetite as well as the stimulation of insulin release

To conclude, bacteria in the GI tract, also called gut flora or microbiome, help with digestion. Parts of the nervous and circulatory systems also play roles in the digestive process. Simultaneously, the nerves, hormones, bacteria, blood, and the organs of the digestive system completes the complex task of digesting the foods and liquids a person consumes each day.

* * *

PART 2

*Parts of the
Digestive System*

Human Digestive Organs

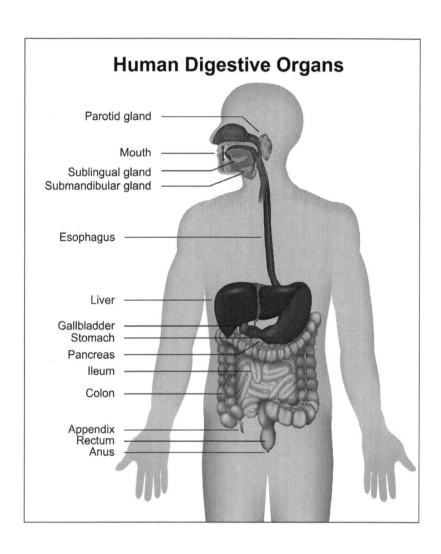

Parotid gland

Mouth

Sublingual gland
Submandibular gland

Esophagus

Liver

Gallbladder
Stomach

Pancreas

Ileum

Colon

Appendix
Rectum
Anus

The human digestive system, (also known as the digestive tract, the GI tract, is a series of connected organs leading from the mouth to the anus. The digestive system which can be up to thirty feet in length in adults, is usually divided into eight parts: the mouth, the esophagus, the stomach, the small intestine, and the large intestine, with the liver, pancreas, and gallbladder. These organs combine to perform six tasks: ingestion, secretion, propulsion, digestion, absorption, and defecation. The digestive system consists of a long passageway called the alimentary canal or digestive tract. Starting at the mouth and ending at the anus.

The parts of the digestive system are:

- The mouth
- The pharynx
- The esophagus
- The stomach
- The small intestine
- The large intestine
- Rectum
- Anus

It's associated organs include:

- The liver
- The pancreas
- The gallbladder

The mouth

The mouth is the beginning of the digestive tract. This is where digestion starts before you even take a bite. Your salivary glands make saliva which is a digestive juice. It moistens food so that it can easily move through the esophagus into your stomach. After you start eating, you chew your food into small pieces that are more easily digested.

Your teeth are also part of the digestive process. Teeth break down food for swallowing and further digestion. The incisors, located in the middle front of the lower and upper jaws, cut and gnaw pieces of food. The molars, in the back of the mouth, grind and chew. The saliva then mixes with the food to begin to break it down into a form your body can absorb and use. The tongue is a muscular organ lying on the floor of the mouth, attached to the mandible and to a small bone called hyoid. It is covered by pink mucous membrane and rough in appearance because it has numerous papillae (the taste buds).

During chewing, the tongue mixes food with saliva, rolls it into a bolus and passes it on to your throat or gullet, and into your esophagus. Your salivary glands make saliva which is a digestive juice responsible for moistening food so that it can easily move through the esophagus into your stomach.

The roles of the mouth in digestion

Teeth breaks down food for swallowing and further digestion with the help of saliva a digestive juice produced by the salivary glands to moisten food.

The pharynx
The pharynx consists of three main divisions. The anterior portion is the nasal pharynx, the back section of the nasal cavity. The nasal pharynx connects to the second region, the oral pharynx, by means of a passage called an isthmus. The oral pharynx begins at the back of the mouth cavity and continues down the throat to the epiglottis, a flap of tissue that covers the air passage to the lungs and that channels food to the esophagus.

The pharynx also allows food to be passed to the esophagus by nasal tubes. The third region is the laryngeal pharynx, which begins at the epiglottis and leads down to the esophagus. Its function is to regulate the passage of air to the lungs and food to the esophagus. As food passes through the mouth into the pharynx, swallowing becomes involuntary.

The roles of the pharynx in digestion
The pharynx caries food to the esophagus and conducts air to and from the trachea or windpipe during respiration.

The esophagus
The esophagus is about 25cm long and extends from the pharynx through the diaphragm and into the stomach. Here food slides down assisted by gravity, however solids are pushed down by peristalsis (a series of muscle contractions that occur in the stomach) The esophagus receives food from your mouth when you swallow. The epiglottis is a small flap that folds over your windpipe as you swallow to prevent you from choking (when food

goes into your windpipe). The esophagus does not produce digestive enzymes, but it does secrete substantial
amounts of mucus that act as a lubricant ensuring the
smooth and easy movement of food.

The stomach

The stomach is a hollow organ that holds food while it
is being mixed with stomach enzymes. These enzymes
continue the process of breaking down food into a usable
form. Cells in the lining of your stomach secrete a strong
acid and enzymes that are responsible for the process
of breaking down the food. The stomach helps the food
with a mixture of these acids and digestive juices to
further break it down. When the contents of the stomach
are processed enough, they are released into the small
intestine.

The stomach lining is protected by mucus from being
damaged by these powerful acids.

Roles of the stomach in digestion

The stomach stores food as well as mixing and grinding it
further down, with the help of gastric juices and enzymes.

The small intestine

The small intestine is divided into three parts; the top
part is called the duodenum. The duodenum is the first
segment of the small intestine. It is responsible for the
continuous breaking-down process. The middle part is
the jejunum and the lower part known as the ileum. The
jejunum and ileum are responsible for the absorption

of nutrients into the bloodstream. Contents of the small intestine start out semi-solid and end up in a liquid form after passing through water, bile, enzymes, and mucus. This contributes to the change in consistency. Once the nutrients have been absorbed and the leftover food residue has passed through this organ, it then moves on to the large intestine. Each part of the small intestine performs a significant role in nutrient absorption.

The duodenum – The chyme first enters the duodenum where it is exposed to secretions that aid digestion. The secretions include bile salts, enzymes, and bicarbonate. The bile salts from the liver help digest fats and fat-soluble vitamins (Vitamin A, D, E, and K). Pancreatic enzymes help digest carbohydrates and fats. Bicarbonate from the pancreas neutralizes the acid from the stomach.

Jejunum – The chyme is then further transited down into the jejunum. Mainly in the first half of the jejunum, the majority (about 90%) of nutrient absorption occurs involving proteins, carbohydrates, vitamins, and minerals.

Ileum – The ileum is the last section of the small intestine and leads to the large intestine or colon. The ileum absorbs water, bile salts, and vitamin B12.

The small intestine roles
It breaks down food and absorbs nutrients such as vitamins, minerals and electrolyte.

The large intestine

The large intestine is the last main site for breakdown and uptake of nutrients. It is made up of the colon, rectum and anal canal or anus. The top part of the large intestine is the colon. By the time the food reaches the colon, many of the useful nutrients have been removed. What is left is watery waste containing a few minerals, vitamins and indigestible materials like fibre. Unlike the small intestine, the large intestine produces no digestive enzymes. Chemical digestion is completed in the small intestine before food residue reaches the large intestine. The food residue is normally brown in color when it reaches the colon. The function of the large intestine is to remove the water and convert the residue into feces. Feces are usually brown in color and contain millions of bacteria. These bacteria are harmless in the bowel.

Functions of the large intestine in digestion

The major functions of the large intestines are to absorb substantial amounts of the water that was added to the food substance all the way along the digestive system. It also houses the Escherichia coli also known as E. coli anaerobic bacteria that metabolize some of what our bodies cannot. They live in symbiosis with us because they can obtain their nutrients from our waste materials. Their metabolism releases minerals and manufactures some vitamins and amino acids. These nutrients get absorbed along with the water into the circulatory system. The bacteria begin the decomposition of the waste materials and convert them to feces with a changed color, smell, and texture.

Acid reduction – The large intestine's mucosa produces bicarbonates which neutralize acidity caused by the synthesis of fatty acids. Furthermore, the large intestines' mucosal layer acts as a barrier, protecting against microbial infections.

The rectum

The rectum is the continuation of the large intestine. Located between the wide passageways between the end of the large intestine and anal canal. It is around 12cm long, and is normally empty, except just before and during defecation. When the feces reach the rectum, the walls are stretched. The sphincters then relax, allowing feces to pass through the anus. Feces are waste products and must be passed out, otherwise if not, toxins build up in the system, which is not healthy or pleasant.

Rectum function
The rectum is the last stop before the feces are eliminated through the anal canal. Similar to the colon, electrolytes are absorbed (sodium, potassium, chloride) and Anaerobic bacteria decompose indigestible food ingredients. The stool is thickened through water absorption and mixed with mucus. The rectum is also part of the continence organ and plays a key role in the mechanism of defecation. If stool enters the rectal ampulla, which is usually empty, it is registered by stretch receptors. The information is transferred to the central nervous system giving the person the urge to defecate. He can now decide to either initiate or delay the defecation by relaxing or tensing the levator-ani muscle and sphincter ani-externus muscle.

However, the rising pressure in the ampulla leads to an increasing relaxation of the involuntary smooth corrugator cutis ani-muscle and sphincter ani-internus muscle (Recto-anal inhibitory reflex) which is why holding in the stool for a long time involves an increasing 'effort'. The rectum supports the defecation through contraction.

The anus

The anus is the last part of the digestive tract. It is a 2-inch-long canal consisting of the pelvic floor muscles and the two anal sphincters (internal and external). The lining of the upper anus can detect rectal contents. And when it does, it lets you know whether the contents are liquid, gas or solid. The anus is surrounded by sphincter muscles that are important in stool control. The pelvic floor muscle creates an angle between the rectum and the anus that stops stool from coming out when it is not supposed to. The internal sphincter is always tight, except when stool enters the rectum. This prevents stools from coming out involuntarily for instance when we are asleep or unaware of the presence of stool. When we get an urge to go to the bathroom, we rely on our external sphincter to hold the stool until reaching a toilet, where it then relaxes to release the stools. Digestive system associated organs are:

The pancreas

The pancreas is a fish shaped greyish-pink gland. It lies in the curve of the duodenum with its body laying posterior to the stomach, and it's tail in contact with the spleen. The pancreas is made up of lobules which secrete pancreatic juice which is then secreted by the pancreatic duct into the duodenum. The juice consists of water, mineral salts, the enzyme trypsin, amylase and lipase. During digestion, these enzymes help to break down sugars, fats and starches.

Functions of the pancreas

The role of the pancreas is to produce pancreatic juices called enzymes that break down fats, sugars, and starches. It produces the hormones insulin and secrets it into the blood stream where it regulates the body's sugar level.

The liver

The liver is the largest gland in the body. It is located in the upper-right hand portion of the abdominal cavity, beneath the diaphragm and on top of the stomach. It is a wedge shaped soft reddish brown in color. The liver makes a digestive juice called bile. Bile helps digest fats and some vitamins. Bile ducts carry bile from the liver to the gallbladder for storage or to the intestines for use. The liver is the most important and active organ in the body.

Roles of the liver

- The liver produces bile which helps carry away waste and break down fats in the small intestine during digestion.
- The liver processes nutrients absorbed from the small intestine.
- Anything that is ingested, from food, drink, alcohol, toxins, gets filtered by the liver The liver is the body's chemical factory, so to speak,

- The liver removes poison from the blood stream such as alcohol and toxins resulting from bacterial infection.
- The liver processes nutrients so they can be utilized by the body and breaks down toxic substances like alcohol, drugs and medicines and removes them by excreting waste into the blood stream which is then cleansed by the kidneys where it's filtered and then removed from the body through urine.
- It converts fat which is stored in the body into a suitable form for combustion
- Excess amino acids are broken down into glucose and urea. Glucose is used as fuel and urea is excreted by the kidneys in the urine. Glucose is stored by the liver in the form of glycogen as a result of the action of insulin.
- It stores iron and vitamins A, B12, D, E and K
- It forms vitamin A from carotene
- It forms plasma proteins which include the clotting substances

The gallbladder
The gallbladder is a pear-shaped organ lying on the posterior surface of the liver. It stores concentrated bile by removing the water. The gallbladder also stores bile between meals. When we eat, the gallbladder squeezes bile through the bile ducts into the small intestine. Bile's most vital role is to break down fats, which are the hardest part of food to digest. When fats are digested, the gallbladder releases bile.

Role of gallbladder in digestion

Fat is difficult to digest, and it resists being broken down into usable energy. This is where bile, comes in to assist breaking them down. As it were, the gallbladder's main function is to store and concentrate bile. Every time a fatty meal is consumed, the gallbladder contracts to release the bile it has stored into the small intestine where it helps the body break down and absorb fats from the food.

* * *

PART 3

The Importance of Nutrition in Digestion

Nutrients encompass all substances that are useful to the body. These include complex chemicals broken down to release energy, carbohydrates, proteins, and fats which are chiefly for building the structural parts of cells: and vitamins and minerals which ensure healthy functioning. The digestive system absorbs the nutrients into the blood stream and lymph at various stages along the digestive tract.

Good nutrition plays a vital role in our lives. To maintain good health and a healthy digestive system, it is important to consume a healthy balanced diet. Together with keeping active, diet and nutrition are crucial factors in;

- Avoiding ill health-preventing illnesses and infection
- Improving mental health – the right diet combined with exercise can help to ease symptoms of depression, stress, and anxiety
- Nutrients are needed for growth and repair
- For energy needed to keep active

Classification of the essential nutrients
1) Macro nutrients
2) Micronutrients

Macro nutrients are:
- Carbohydrates
- Proteins
- Fats

Micronutrients are:
- Vitamins
- Minerals
- Other – fibre, water, light, oxygen

Carbohydrates

Carbohydrates are the major fuel of the body. Plants make carbohydrates by combining the sun's energy with hydrogen and oxygen in the water drawn up by the root system, with carbon and more oxygen from the carbon dioxide in the air. When we eat the plant, our digestive system combines it with oxygen and the sun's stored energy it released to enable us to move and think.

Carbohydrates are grouped together as sugars and starches are classified as to how many sugar units are present in the molecule. The different types of carbohydrates include monosaccharides, disaccharides or polysaccharides.

Monosaccharides are single units of sugar for example:
- Glucose-the body's main source of energy
- Galactose-which is readily available in milk and dairy products
- Fructose-mostly occurs in fruit and vegetables

Disaccharides are two sugar molecules joined together, for example:
- Lactose-found in milk which is made up of glucose and galactose
- Sucrose or table sugar is made up of glucose and fructose

Polysaccharides are chains of many sugars. They can consist of hundreds or thousands of monosaccharides. Polysaccharides function as food stores for plants and animals.

Examples include:
- Glycogen – stores energy in the liver and muscles
- Starches-abundant in potatoes, rice, and wheat
- Cellulose-is one of the main structural components of plants.

Carbohydrates are found in two forms:
a) Slow releasing form: This is contained in whole grains, vegetables, and fresh fruit which contain mere complex carbohydrates and more fibre. When consumed, these foods slow down the release of the sugar into the blood stream.
b) Fast releasing form: This is contained in sugar, honey, sweets, refined flour and foods. These products send a sudden rush of energy in the blood stream, this energy requires a certain amount of insulin to balance it. When the body expends energy fast, there is an immediate need of another fix, therefore the cycle goes on and forth, often spiraling into addiction which is dangerous and unhealthy.

How does the digestive system break down carbohydrates?

After the digestive system processes glucose or blood sugar, the blood stream absorbs glucose and uses to as energy to fuel the body. The amount of carbohydrates consumed affects blood sugar. For instance, taking a vast sum of carbohydrates can raise blood sugar levels which can put you at risk of diabetes (hyperglycemia) On the other hand, people who consume less carbohydrates are at risk of low blood sugar(hypoglycemia)

Carbohydrates are found in:
- Whole grain products such as brown rice, oatmeal, whole wheat bread and pasta
- Beans and legumes eggs black beans, kidney beans, chickpeas, lentils, peas
- Vegetables – broccoli, Brussels sprouts, carrots, potatoes, green beans
- Fruit – apples, mango, pineapple, berries, bananas, melons, raisins.

Importance of carbohydrates
1) Carbohydrates are the body's main source of energy for the most obvious functions of the body like thinking and movement.
2) Helps preserve muscle
3) They promote digestive health – fibre helps ensure good gut function increasing the physical bulk in the bowel, thereby stimulating the intestinal transit.
4) They boost heart health

Proteins
Proteins are large group molecules that perform many functions in the body. Proteins are of great nutritional

value and are directly involved in the chemical processes essential for life. Chemically, protein is composed of amino acids which are organic compounds made of carbon, hydrogen, nitrogen, oxygen, or sulfur. Protein is found throughout the body; in the hair, skin, bone, muscle, and tissue. Some proteins known as complete proteins contain all the essential amino acids.

Proteins obtained in the diet are from animal sources are complete. Those obtained from plant sources are incomplete-which means that they are deficient in one or more essential amino acids. To get all the essential amino acids from plant sources, it is necessary to combine different plant sources for example beans and corn.

Examples of proteins are:
- Meat – lean beef, lean pork
- Fish
- Poultry – chicken, turkey, eggs, duck, goose
- Milk
- Nuts and seeds
- Beans and pulses
- Peas and chickpeas
- Cheese
- Oats
- Corn
- Yoghurt
- Soy products
- Avocado
- Potatoes
- Vegetables – broccoli, cauliflower, Brussels sprouts, asparagus

How does the digestive system break down proteins?
Proteins are digested in the stomach and small intestine. In the small intestine, protein is broken down into peptides which are then broken down into single amino acids that are absorbed in the blood stream, delivering them to cells in other parts of the body so they can start repairing tissue and building muscle.

Importance of proteins

1) For repairs and maintenance. Protein is the building block of the body. It is essential in the maintenance of body tissue, for development and repair. Our organs, hair, skin, eyes, and muscles are all made from protein.
2) For energy – protein is the main source of energy in the body.
3) Enzymes – these are proteins that increase the rate of the chemical reactions in the body.
4) Hormones – protein engages in the making of some hormones.
5) Antibodies – these are protective proteins produced by the immune system in response to presence of a foreign substance in the body. Therefore, helpful in preventing infection in the body.
6) Proteins aid in the transportation of some molecules for example hemoglobin which is a protein that transports oxygen throughout the body.
7) Proteins provide structure. Because they are fibrous in nature, they give cells and tissues the stiffness

and rigidity. Such proteins are keratin, collagen, and elastin.

8) Proteins play a crucial role in regulating the concentration of acid and bases in blood and other bodily fluids.

9) They regulate body processes to maintain fluid balance for example albumin and globulin protein.

Fats

Fat is a source of essential fatty acids, which the body cannot make itself. A small amount of fat is an essential part of a healthy balanced diet as it helps the body absorb vitamins A, D and E. These vitamins are fat-soluble, which means they can only be absorbed with the help of fats. Any fat that is not used by your body's cells or turned into energy is converted into body fat. The same applies to unused carbohydrates and proteins, they too are converted into body fat.

All types of fat are high in energy. A gram of fat, whether it is saturated or unsaturated, provides 9kcal (37kJ) of energy compared with 4kcal (17kJ) for carbohydrate and protein. Akin to protein and carbohydrates, fat is a type of nutrient which the body needs to function.

Types of fats

1) Unsaturated fats – ("good fats") Unsaturated fats are found in foods from plants such as nuts, seeds, vegetables oil. There are two types of unsaturated fats:

a) Mono-saturated fats, found in:
   - Avocados
   - Olive oil, olive oil spreads
   - Nuts – hazel nuts, almonds, peanuts
   - Seeds

b) Poly-saturated fats – found in:
   - Fish
   - Walnuts
   - Sunflower
   - Soybeans
   - Flax
   - Corn
   - Canola oil

2) Saturated fats – ("bad fats") Saturated fats are found in animal foods; however, some plant foods are also high in saturated fats like coconut and palm oils.

Foods high in saturated fats include:
   - Meat products
   - Fatty cuts of meat
   - Butter, ghee, and lard
   - Palm oil
   - Coconut oil
   - Cakes, pastries, biscuits
   - Cheddar cheese
   - Chocolate confectionery
   - Cream, ice cream

Cholesterol and saturated fats

Cholesterol is a fatty substance that is mostly made by the body in the liver. Eating too many fats in the diet can raise cholesterol levels in your blood which in turn increases the risk of stroke and heart disease. It is recommended that healthy adults should have a total cholesterol level below 5 mmol/L The total cholesterol level includes LDL (bad cholesterol) and HDL (good cholesterol). The risk of coronary heart disease is particularly high if you have a prominent level of LDL cholesterol and a low level of HDL cholesterol. Individual levels of LDL and HDL cholesterol will vary, and your doctor will be able to give you specific advice based on your own results.

How does the body digest fats?

When we eat fats, stomach emptying slows down. The digestion of fats begins in the mouth using enzymes in the saliva, these are: lipase and phospholipids. Digestion continues to the stomach, but more of it happens in the small intestine. In the stomach, gastric lipase further breaks down the fats. As the stomach contracts with intensity, the fats are converted into diglycerides and fatty acids. The contents of the stomach including diglycerides and fatty acids travel to the small intestine.

The liver releases bile which contains lecithin, bile salts and emulsifiers that help further break down fats. Once fat has been broken down into water form, it is released into the blood stream and carried to tissues that require energy. The hardest fats to digest are the solid fats – those that are solid at room temperature for example butter, cheese, and lard.

How to improve fat digestion

– Eating more healthy fats such as nuts, seeds, avocados, fish oils and reducing the intake of processed fats.
– Eating a low-fat diet if possible

Functions of fats in the body

– For energy – fat can be stored in the body's fatty tissue which releases fatty acids when energy is required.
– Fats carry vitamins A, D, E and K and supports their absorption in the small intestine.
– Fats provide protection and insulation. Fat stored within the abdominal cavity (subcutaneous fat) and underneath the skin (visceral fat) protects the vital organs such as the heart, liver, kidneys. The subcutaneous fats insulate the body from extreme cold temperatures and helps keep the internal climate under control.
– Fats help the body to produce and regulate hormones.
– They support cell growth
– Fats keep cholesterol and blood pressure under control.
– Fats give shape to the body

Micronutrients

Vitamins

Vitamins are organic nutrients found in foods and are essential in small quantities for both growth and good health. Chemically, vitamins are made from the same elements: carbon, hydrogen, oxygen and sometimes nitrogen. But their elements are arranged differently and therefore perform distinct roles in the body.

The human body needs small amounts of vitamins and very small amounts are available in foods. Most vitamins are obtained through food, but the bacteria in the intestines produce a few. Part of the required vitamin D is produced in the skin when it gets exposed to the sunlight. To get all the vitamins our bodies need, we must eat a variety of foods as no single food will supply them all. Because vitamins do not have any calories, they do not directly provide energy in the body, but they are involved in energy metabolism.

Vitamins are classified according to how soluble they are, in either fat or water form. The fat-soluble vitamins are A, D, E and K. These occur in foods containing fat, and they can be stored in the body. The water-soluble vitamins are vitamin C and B complex. They are not stored in the body.

Vitamin A food sources:
- Eggs
- Cheese
- Oily fish – salmon, mackerel, herring
- Milk and yogurt
- Liver
- Sweet potato
- Spinach
- Fortified low-fat spreads
- Carrots

Vitamin B types and food sources:

Thiamine (B1) – Even though thiamine is found in every body tissue, the body does not store it, hence the need for a daily intake. Food sources are nuts, whole grain, peas, liver, some fortified breakfast cereals, some fresh fruit, and vegetables. turn carbohydrates and sugar into energy. Thiamine plays a key role in nerve, muscle, and heart function, and is essential for a healthy brain.

Riboflavin (B2) – Like thiamine, this vitamin is coenzyme and without it, the body cannot digest or use proteins and carbohydrates. Sources found in milk, eggs, mushrooms, plain yogurt, fortified breakfast cereals. Riboflavin is important for energy release, helps the body to absorb nutrients, including iron, and is crucial for healthy eyes, skin and nervous system.

Niacin (B3) – found in meat, fish, eggs, wheat flour. Niacin Helps release energy from food, keeps the nervous and digestive systems healthy, and is essential for normal growth and healthy skin.

Pantothenic acid – Vitamin B5 – found in beef, chicken, eggs, liver, and kidney, mushrooms, avocado. Pantothenic helps the body turn protein and fat into energy, and is also vital to enzyme reactions that make it possible for the body to use carbohydrates and create steroid bio-chemicals such as hormones.

Pyridoxine – Vitamin B6 is a component of enzymes that metabolizes proteins and fat. It's found in pork, chicken, turkey, fish, lamb, peanuts, some fish, oats, soybeans, milk, banana, wheat germ. Vitamin B6 breaks down protein from food, reduces tiredness and fatigue, and is essential for healthy red blood cells and nervous system.

Biotin (B7) – this vitamin is needed in tiny amounts. Bacteria that naturally live in the bowel can make biotin, also found in a wide range of foods, with extremely low levels in egg yolks, legumes, nuts and seeds, liver, sweet potato.

Biotin helps the body break down nutrients from food and process glucose, maintain mucous membranes and keeps skin and nails healthy.

Folate and folic acid (B9) – found in fortified breakfast cereals, kidney beans, peas, chickpeas, liver, and leafy green vegetables such as spinach, kale, spring beans, cabbage, broccoli, Brussels sprouts.

This vitamin is essential during pregnancy to help prevent neural tube defects (spina bifida) in babies, helps maintain healthy red blood cells, and contributes to reduction of fatigue.

Vitamin (B12) – found in meat, fish, milk, cheese, eggs, beef, liver. It is vital in the nervous and immune systems, helps fight tiredness and fatigue, and needed for healthy DNA.

Vitamin C food sources:
- Citrus fruits – oranges, lemons, lime, grapefruit
- Broccoli
- Potatoes

- Brussels sprouts
- Kale
- Kiwis
- Mango
- Papaya
- Pineapple
- Bell peppers
- Strawberries
- Tomatoes
- Cauliflower

Vitamin functions

- Maintaining a healthy skin, blood vessels and bone cartilage
- Helping with wound healing
- Boosting the immune system
- Aid in maintaining a healthy nervous system

Vitamin D sources:

- Oily fish – salmon, sardines, herring
- Red meat
- Liver
- Egg yolks
- Fortified foods – cow's milk, orange juice, breakfast cereals
- Yogurt

Function of vitamin D

- Help the body to absorb and retain calcium and phosphorus which are essential for bone building

- Keeps muscles healthy
- Vitamin D helps regulate calcium and phosphorus in the body, thereby keeping bone, teeth, and muscles healthy.

Vitamin E food sources:

- Nuts and seeds
- Plant oils – olive oil, vegetable, oil, sunflower oil canola oil, peanut oil
- Wheat germ – found in cereals and cereal products, avocado, fish

Functions of vitamin E

- Helps maintain healthy skin and eyes and strengthening the body's natural defense against illnesses and infection.
- Helps maintain healthy heart function
- Good for cognitive health

Vitamin K sources:

- Leafy green vegetables
- Vegetable oils
- Cereal grains
- Liver
- Avocado
- Soft cheese

Uses of vitamin K

- Useful for blood clotting-helping in would healing
- Helps improve bone health

How does the digestive system break down vitamins?
Vitamin digestion depends on what type of vitamin had been ingested by the body, for instance if it is fat soluble or water soluble like proteins, fats, and carbohydrates. The digestion of vitamins is very much standardized. However, it is in the small intestine that the vitamin absorption happens. Water soluble vitamins such as vitamin C and B are picked up in the middle part of the small intestine called the jejunum. The molecules then transport them through wall cells of the intestine and deposits them in the body where they enter the blood stream. Because water soluble vitamins dissolve in water, they do not require stomach acids for absorption.

On the other hand, fat soluble vitamins like A, D, E and K need to dissolve in fat before they can make way into the body. They require fat digesting bile acids from the liver. After the bile acids break down the fat, the vitamins are dissolved in, the vitamins then move with the fat through the intestinal wall into the body and finally end up in the liver where they are needed. Whereas water soluble vitamins require daily consumption, fat soluble vitamins do not, as the body stores them.

Minerals
To function well, the body needs a number of minerals. We need some minerals in daily quantities of over 100 mg and these we call macronutrients. They include calcium, magnesium, phosphorus, potassium, and sodium. The other group of minerals are micro-minerals, they are

required in smaller amounts below 100 mg. They include iron, chromium, copper, manganese, and zinc.

Macro minerals
Calcium and phosphorus
These are used for building bones and teeth, giving rigidity to the structures. Calcium circulates in the blood and appears in other body tissues where it helps blood to clot, muscles to contract including the heart muscle and helps the nerves to transmit impulses.

Food sources of calcium are:
- Milk or milk products
- Canned fish, for example sardines, salmon eaten with bone
- Certain shellfish
- Several greens like spinach, broccoli
- Dried beans and peas

Phosphorus engages in the release of energy from fat, protein, and carbohydrates during metabolism and in formation of DNA and many enzymes. Phosphorus plays a major role in balancing the acids and alkalis produced in the body. Phosphorus can be found in:
- Milk products
- Fish
- Eggs
- Legumes
- Meat

- Whole grain foods
- Nuts

Sodium, potassium and chloride

These minerals are collectively known as electrolytes because when dissolved in the body fluids, they separate into positively or negatively charged particles called irons.

Sodium is mainly found in the fluids outside the cells (positive charge)

Potassium is found mainly within the cells (with a positive charge).

Chloride is also found mainly outside the cells.

These three electrolytes maintain the water balance by moving the water around in the body. They also have the capacity to neutralize various acids and alkalis in the body.

Sodium is needed for muscles to contract and nerve impulses to be transmitted. Table salt is the major source of sodium in the diet. Sodium can also be found in baking powder and soda.

Potassium assists in muscle contraction, including maintaining a normal heartbeat and sending nerve impulses. Sources of potassium are both animal and plant based like meats, poultry, fish, milk, yogurt, legumes, potatoes, soybeans.

Chloride is essential for nerve function and is also part of hydrochloric acid which is found in high concentrations in the juices of the stomach. Sources include:

- table salt or sea salt
- celery
- tomatoes
- seaweed
- lettuce
- rye.

Magnesium is a mineral found in all body tissue, bone and in the blood. It is a vital part of many enzyme systems responsible for energy conversions in the body. Magnesium also has a role in making protein and maintaining a normal basal metabolic rate. Sources are: green leafy vegetables, seeds, whole grain cereals, nuts, legumes, seafood.

Sulfure – sulfure is found in some of the amino acids found in protein and also in the vitamins biotin and thiamine. The protein in hair, skin and nails is particularly rich in sulfure. Sources are: eggs, chicken, turkey, duck, fish, nuts, seeds, whole grains, legumes, leafy greens

Iron is needed for growth and development. The body also uses iron to make hemoglobin a protein in red blood cells that carries oxygen from the lungs to all parts of the body. Iron is also needed in hormone production. Sources are: red meat, liver, kidney beans, edamame beans, chickpeas, nuts, dried fruit, fortified breakfast cereals.

Zinc is needed for growth and maintenance of the human body. It is found in several systems and biological reactions. Zinc is also needed for immune function, blood clotting, thyroid function. Sources are: meat, sea foods, diary-products, nuts, seeds, legumes, whole grains.

Digestion of minerals

The absorption of minerals differs according to the mineral and other nutrients eaten at the same time. Minerals in animal foods tend to be absorbed better than those in plant foods. However though, many minerals are absorbed in the small intestine.

Fibre

Fibre is a carbohydrate found in whole grain cereals, fruits, and vegetables. Fibre is very important for keeping the digestive tract working smoothly. It is made up of the indigestible parts or compounds of plants, therefore, passes relatively unchanged through the stomach and intestines. Its main role is to keep the digestive system healthy. Fibre is also termed as roughage or bulk. Dietary fibre is divided into two categories: soluble fibre and insoluble fibre.

Soluble fibre

Soluble fibre easily dissolves in water and is broken down into gel like substances in the part of the small intestine and then continues to the large intestine.

Examples of soluble fibres are:
- Oats
- Nuts
- Beans
- Fruit – apples, blueberries, strawberries, citrus fruits
- Cereals
- Rice

- Barley
- Seeds
- Seaweed

Insoluble fibre does not dissolve in water or gastro-intestinal fluids, it remains majorly unchanged as it moves through the digestive tract. It's found in:

- Leafy greens like cabbage, spring greens, cauliflower
- Nuts
- Carrots
- Parsnips
- Potatoes
- Bran
- Whole wheat
- Rye
- Beans
- Peas

The benefits of fibre in the diet
Soluble fibre:

- lowers cholesterol by preventing some dietary cholesterol from being broken down and digested.
- lowers fat absorption, hence helping in weight loss and management.
- stabilizes blood sugar levels as it prevents fats from being broken down and digested.
- Soluble fibre feeds healthy gut bacteria and aids in their longevity in the large intestine.
- reduces the risk of cardiovascular disease by lowering cholesterol levels.

Insoluble fibre benefits

- It prevents constipation
- It helps maintain bowel health
- It stabilizes bowel movements by softening the stools so that they're easier to pass.

How fibre is digested

Unlike other food components like proteins, carbohydrates, and fats which the body breaks down and absorbs, fibre passes through the stomach to the small intestine, then proceeds to the large intestine and out of the body. Soluble fibre soaks up water as it goes through the digestive tract, this helps bulk up the stool and prevents constipation and diarrhea. A diet low in fibre may cause constipation, on the contrary, a diet with too much fibre may cause diarrhea. That's why it's very important to strike the right balance.

How much fibre do you need?

Government guidelines published in July 2015 suggest that our dietary fibre intake should increase to 30g a day, as part of a healthy balanced diet. Children under the age of 16 don't need as much fibre in their diet as older teenagers and adults.

2- to 5-year-olds need about 15g of fibre per day, 5- and 11-year-old need about 20g, 11- to 16-year-olds need about 25g of the recommended dietary reference value for adults is 18 grams a day.

If you eat more fibre than you need, your body will tell you by protesting with diarrhea, gas or in extreme cases cause intestinal obstruction. This is because too much roughage irritates your intestinal tract

Water
Water is one of the main nutrients of the body. Arguably, it's the main nutrient because the body is approximately 60–70% water. Water is a solvent, it dissolves other substances and carries nutrients and other material around the body, enabling every organ to work efficiently.

The importance of water in digestion
It is crucial to be mindful that every day, the body loses water through breathing, sweating, urine, bowel movement. That being the case, it is very important to keep it hydrated not only for a healthy and nicely well-watered digestive system, but also for your general well-being. Unlike other nutrients, the body doesn't store water, so you must take in a new supply every day – enough to replace what you lose. This can be achieved through tap water, spring bottled water, soft drinks, fruit juices and smoothies, milk, tea, and coffee.

How water supports a healthy digestive system
From the minute the food goes into the mouth, water helps break it down in form of saliva and other digestive juices and enzymes, and allows nutrients to be absorbed in the body. Both small and large intestines absorb water as it moves into the blood stream where it is also used to break down nutrients. As the large intestine absorbs water, stool changes from liquid to solid.

Dehydration is one of the most common causes of constipation, that's because when you do not drink enough water, your body becomes dehydrated. The large intestine soaks up the water from our food waste, which results in a dry hard stool that is not only difficult to pass, but also painful. Drinking water regularly will result in the food waste passing through the intestine easily and on to the colon and eventually vacated as faeces.

Chewing and swallowing are as natural as breathing in our lives, but we rarely give both processes a second thought throughout the day. These processes can run efficiently because of the saliva that is ever present and on standby in the mouth. The primary constituent of saliva is water and decreased body water can lead to decreased saliva production, causing bad breath and the sensation of food getting stuck the throat.

Signs of dehydration
- Constipation
- Reflux
- Peeing less
- Concentrated dark yellowish strong-smelling urine
- Headache
- Feeling thirsty
- Dry mouth, lips, and eyes
- Nausea or dizziness
- Impaired coordination
- Fatigue and irritability
- Confusion
- Reduced muscle strength

* * *

PART 4

Diseases of the
Digestive System

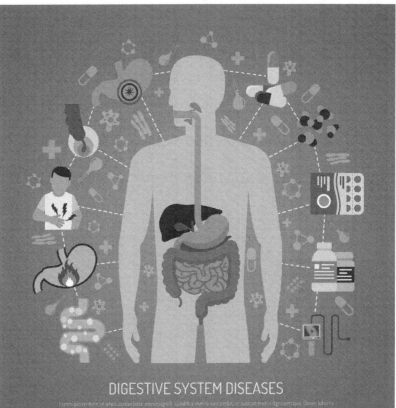

DIGESTIVE SYSTEM DISEASES

Lorem ipsum dolor sit amet, consectetur adipiscing elit. Curabitur viverra sem lacinia, in suscipit metus dignissim quis. Donec lobortis consectetur gravida. Aliquam sollicitudin ligula, quis quis elementum ligula. Nullam mattis bona non fermentum consequat, justo nec sollicitudin turbo, at orci porttitor nisant mi lorem. Vivamus id sollicitudin mi.

The digestive system plays a major role in our overall health, as it is the channel for nutrients our bodies need for survival. Without good digestive health, you will experience a host of uncomfortable conditions both short-term and lasting.

Gastrointestinal diseases affect the gastrointestinal (GI) tract, from the mouth to the anus. There are two types of conditions: functional and structural. Structural conditions are those where your bowel looks abnormal upon examination and doesn't work properly, these include colon cancer, colon polyps, inflammatory bowel disease, diverticular disease and hemorrhoids. Functional conditions or illnesses are those whereby the GI tract looks normal when examined but doesn't move properly. Examples are:

- constipation
- irritable bowel syndrome or IBS
- diarrhea
- food poisoning
- GERD (Gastroesophageal Reflux Disease).

Constipation

This functional problem is caused by infrequent bowel movement when it's difficult to pass stools. Often when people are constipated, they strain during bowel movement and it can be a painful and uncomfortable experience. It may cause dry hard stools and anal injuries like fissures and hemorrhage. Constipation is caused by:

- eating a diet low in fibre
- not drinking enough water
- lack of exercise,
- eating a lot of diary-products
- overuse of laxatives
- certain medicines
- stress, anxiety
- depression
- ignoring the urge to go the toilet
- changing diet
- changes in hormones for instance during pregnancy; and after childbirth.

Treating constipation

- Drink plenty of fluids throughout the day
- Making sure your diet has enough fibre in it as it promotes regular bowel movement
- Keep active – exercise can help prevent and also relieve constipation
- Do not put off pooping – go as often as your body tells you too

Diarrhea

Like constipation, diarrhea is a very common condition that affects almost everyone at some point in their lifetime. Diarrhea is when you have loose and watery stools. People who have diarrhea more often than normal also have IBS. Diarrhea usually happens when a virus or bacteria gets into the lining of the intestines, causing irritation and strong contractions. When this happens, a large quantity of fluid is pared through the blood stream into the bowel, thus, the bowel moving its contents rapidly than normal.

Acute diarrhea can last several days and clear itself or can be treated with over-the counter medicines, though, chronic diarrhea can last longer. This may be because of IBS, Crohn's disease, ulcerative colitis, or any other illnesses where food is not absorbed properly. Food intolerance such as lactose intolerance can also irritate the bowels and cause diarrhea. Acute diarrhea should be reported to the doctor for further diagnosis or treatment, as it can cause severe dehydration.

Causes of diarrhea
- Eating foods that upset the digestive system such as too much fibre
- Allergic and intolerance to certain foods like milk products
- Medications
- Bacteria
- Infections and other organisms

Certain fibre foods that can help make stool more solid are:
- Bananas
- Potatoes
- white
- rice
- white bread
- noodles
- chicken
- turkey

- lean beef
- eggs

Irritable Bowel Syndrome (IBS)

IBS is one of the most common digestive system conditions. IBS is sometimes described as "functional bowel disorder" because it affects the way the large intestine works. IBS is usually a lifelong problem and can be frustrating to live with and can also have a big impact on one's everyday life. People with IBS often have irregular colon motility patterns because the necessary muscle contractions are not functioning the way they should. The term "irritable" is used because the nerve endings in the lining of the bowel are unusually sensitive, and the nerves that control the muscles of the gut are unusually active.

IBS is diagnosed based on your symptoms and elimination of other causes. Your doctor will take a detailed medical history and perform a thorough physical exam. Unlike IBD, IBS cannot be confirmed by visual examination or with diagnostic tools and procedures, though your doctor may use blood and stool tests, x-ray, endoscopy, and psychological tests to rule out other diseases. There is no cure for IBS, but dietary changes and medicines can often help control the symptoms.

Symptoms of IBS

- – Changes in bowel habits – experiencing constipation or diarrhea
- – Passing excess gas

- Abdominal pain and cramping which often reduce after passing stool
- Feeling as though the bowels are not empty after passing stools
- Passing of mucus from the rectum
- A sudden, urgent need to go to the toilet
- Swelling or bloating of the abdomen

IBS symptoms often tend to get worse after a meal. Symptoms vary between people.

Although some people have mild symptoms, they can be severe in others. The exact cause of IBS is not known, but as hinted earlier on, certain things trigger the symptoms of IBS, these include food, medications, and stress. Besides medication, many people find that dietary changes improve their symptoms. For example:

- Adding or increasing fibre in your diet – eating more fruit, vegetables, nuts, grains, seeds
- Drinking plenty of water
- Limiting cheese and milk as lactose intolerance is common in people with IBS
- Regular exercise is important as exercise is thought to ease IBS symptoms by minimizing stress, improving bowel function, and reducing bloating.
- Eat smaller meals
- Seek help from a nutritionist if you need further advice on your diet
- Don't smoke
- Avoid fatty foods

- Stay away from gassy foods like beans, lentils, carbonated drinks, as these can trigger IBS
- Caffeinated drinks like coffee, sodas, and energy drinks can all set off IBS
- Avoid sugar free products such as gums, diet drinks
- Chocolate bars and drinks activate IBS
- Limiting alcohol may help reduce the symptoms

Heartburn

Heartburn is a burning feeling in the chest caused by the acid traveling upwards the throat (acid reflux) It can be very uncomfortable, but it is usually harmless. And most people manage the discomfort of heartburn on their own with over the counter medications, lifestyle adjustments or home remedies.

Causes of heartburn

When you eat food, it passes down a long tube that links it to your mouth and stomach. This tube is called the esophagus. At the bottom of the esophagus is a valve called the esophageal sphincter. This value opens to let food through and then closes to keep your stomach contents down. Inside your stomach is a very acidic mixture that starts the process of breaking down the food you have eaten. Your stomach is designed to hold this mixture, however, your esophagus can't hold this mixture without consequences.

In instances where the valve that separates your stomach and esophagus doesn't close properly, some of the acidic mixture from your stomach goes back up the esophagus.

This is called reflux. When you have reflux, you will often feel a burning sensation. That is heartburn. Other conditions like GERD, pregnancy, hiatal hernia and certain medicines can also cause reflux.

Foods that can contribute to heartburn

- Onions
- Tomatoes and tomato products
- High fat foods
- Citrus fruits and juices
- Alcohol
- Caffeinated beverages
- Carbonated drinks
- Spicy foods
- Chocolate
- Peppermint

Triggers of heartburn

- Smoking – nicotine relaxes the lower esophageal sphincter increasing the reflex acids moving up the stomach
- Alcohol – Some people can avoid heartburn by drinking alcohol in moderation, which means one or two drinks only. For others, heartburn can arise even with a small serving of alcohol. Too, some drinks can cause heartburn more than others. Work out which types of drinks trigger your heartburn and avoid them.
- Eating heavy meals
- Eating too fast

- Eating while laying down
- Eating too close to bedtime
- Some medicines for high blood pressure, heart conditions, antibiotics, nitrates can trigger heartburn
- Being overweight
- Wearing tight fitting clothing
- Pregnancy may also relax the muscles and trigger heartburn

Symptoms of heartburn
- A burning sensation in the chest or throat after eating
- A hot sour or salty tasting fluid in the back of the throat
- Difficulty swallowing
- Pressure behind the breastbone
- Needing to burp
- Feeling sick and irritated

You can ease heartburn by:
- Eating smaller meals frequently throughout the day and taking your time to chew it. Also, eat a couple hours before you go to bed.
- Try and lose weight if you are obese or overweight
- Quit smoking
- Wear loose fitting comfortable clothing to prevent pressure on your abdomen
- Avoid late night snacking
- Limit your alcohol intake – beers, wine, distilled spirits all trigger heartburn
- Keep your stress levels in check

- Monitor your medication and check with your doctor how they can help with your condition in relation to your triggers

Indigestion

Indigestion can be a pain or discomfort in your upper abdomen. It may be caused by stomach acid encountering the sensitive, protective lining of the digestive system. The stomach acid breaks down the lining, leading to irritation and inflammation, which can be painful. Indigestion is a chronic disease that usually lasts years, if not a lifetime. It does, however, display periodicity, which means that the symptoms may be more frequent or severe for days, weeks, or months and then less frequent or severe for days, weeks, or months. The reasons for these fluctuations are unknown.

Most people with indigestion don't have inflammation in their digestive system. Therefore, their symptoms are thought to be caused by increased sensitivity of the mucosa. Indigestion does not typically lead to severe complications, however, severe or persistent symptoms may make it more difficult for you to eat the necessary amount of food. This may influence the overall nutritional balance of your diet.

Most people experience indigestion at some point, usually, it's not a sign of anything serious and you can treat it yourself. In most cases indigestion is related to eating, although it can be triggered by other factors such as smoking, drinking, alcohol, pregnancy, stress or taking certain medications.

Symptoms of indigestion

Symptoms of indigestion can vary greatly from one individual to another and depend on the underlying problem causing the indigestion. Most indigestion sufferers have their own pattern of symptoms, which range from mild discomfort in the upper part of the abdomen to quite severe pain, which sometimes may go through into the back. Some people may feel a burning sensation rising in the chest called heartburn, while others experience a more general feeling of fullness and discomfort in the upper abdomen after a meal. Sometimes, a more localized painful sensation just below the breastbone is felt or a combination of all three. Indigestion can occur by itself or may be accompanied by other symptoms such as:

- Upper abdominal pain or discomfort
- Nausea
- Abdominal bloating
- Feeling full after eating only a small amount of food
- Excess burping
- Occasional vomiting
- A burning sensation in the stomach
- A rumbling or gurgling stomach

Causes

When you eat, your stomach produces acid. Indigestion is often caused by excess stomach acid meeting the sensitive, protective lining of the stomach, the top part of the bowel, or the esophagus, which can cause soreness and swelling.

This produces a feeling of discomfort which can feel like a 'fire in your belly', particularly if your digestive lining is especially sensitive to acid.

Treatment for indigestion usually starts by looking for the root causes of the problem. Treatment can also easily be remedied by simple lifestyle changes or exclusion from diet. Most sufferers of indigestion can control their symptoms simply by taking over-the-counter antacid tablets or liquids from the pharmacist. Some of these treatments work by neutralizing stomach acid while others reduce the amount of acid your stomach produces. Your doctor will also be able to advise you about whether any other medication that you are taking is likely to be causing indigestion and prescribe longer-term antacid or acid suppressing medication for you.

Typical causes of indigestion are:

- Diet – people often associate their symptoms with specific foods. Eating heavy meals or fatty foods can irritate the belly. Eating too fast also encourages indigestion. Although specific foods might worsen the symptoms of indigestion, they usually are not the cause of indigestion.
- Smoking – The chemicals in the smoke can cause the muscle between the esophagus and stomach to relax, allowing acid to escape into the digestive system.
- Alcohol – alcohol causes the stomach to produce more acid than normal which can consequently lead to irritation of the lining of the stomach.

- Medication – NSAIDs medications such as ibuprofen and aspirin if regularly taken can affect the digestive tract.
- Being obese or overweight can increase the risk of indigestion as there is increased pressure in the stomach especially after having a meal.
- Pregnancy can have the same effect as above, with the growing baby pressing on the stomach in the last trimester.
- Stress and anxiety – even though stress and anxiety don't cause indigestion, however when they become a vicious cycle, it can exacerbate indigestion and other gastrointestinal conditions.

Crohn's disease

Crohn's disease is a life-long condition where parts of the digestive system become inflamed. Crohn's disease is a type of inflammatory bowel disease. However, it commonly occurs in the small intestine and the large intestine. Crohn's Disease causes ulceration and inflammation, which affects the body's ability to digest food, absorb nutrients and eliminate waste, in a healthy way. Crohn's Disease is one of the two main forms of Inflammatory Bowel Disease (IBD).

The other main form of IBD is a condition known as Ulcerative Colitis. Crohn's is sometimes described as a chronic condition because it's an ongoing and life-long illness, although the sufferer may have periods of good health as well as times when symptoms are more active. In many people, the disease runs a benign course with few flare-ups, while other people may have more severe disease.

In addition, in some people, Crohn's disease results in patches of bowel inflammation with groups of small sores (aphthous ulcers), which are like mouth ulcers. In moderate or severe cases, ulcers become large and deep with a lot of nearby inflammation. Inflammation causes redness, swelling and pain and can make the bowel wall thicken up, narrowing the width of the gut which can block the passage of food. Sometimes, deep ulcers break through the bowel wall causing infection outside the bowel (an abscess). Sometimes, if the infection doesn't heal, or it heals but leaves a channel, a tract called a fistula can open from the bowel to other areas. Fistulas are more frequent around the anus. Also, as healing happens, scar tissue may form, which can in some people lead to an obstruction in the bowel (stricture).

Any part of the gut from the mouth to the anus can be affected. The most common area is the end of the small bowel (terminal ileum) through to the large bowel (colon), near the appendix. In some people, only the large bowel is affected. In others, many parts of the gut are affected and, rarely, the mouth, gullet or stomach. Sometimes, the gut inflammation can also trigger inflammation outside the bowel leading to arthritis and swollen joints, sore red eyes due to inflammation, or skin complaints (rashes). These are known as 'extra intestinal' symptoms.

The range of severity for Crohn's disease is mild to debilitating, symptoms may vary and change over time, but in severe cases of Crohn's disease can lead to life threatening complications. Crohn's disease is not infectious. The symptoms develop gradually and may also change over time, with periods of good health when you have few or no symptoms, alternating with times when your symptoms worsen, with more flare-ups.

Crohn's is an individual condition, whereas some people may remain well for a long time, even for many years, others may have more frequent flare-ups, but the earliest symptoms to look out for include:

- Abdominal pain or cramps
- Diarrhea
- Tiredness and fatigue
- A fever or generally feeling unwell
- Mouth ulcers
- Loss of appetite and weight loss
- Urgent need to open your bowels
- Anemia in suffers that experience blood loss in their stools
- Delayed growth and development in children.

In more severe cases, Crohn's disease can lead to serious complications namely:

- Fissures – these are tears in the lining of the anus and can cause pain and bleeding especially during bowel movements.
- Fistula – these are caused by inflammation in an abdominal channel that forms between one part of the intestine and another, or between the intestine and the bladder, vagina or skin. Fistula s are more

common in anal area and require immediate medical attention.

- A stricture – this is a narrowing of the intestine because of chronic inflammation.

What causes Crohn's disease?

Although there has been much research, it's still not known exactly what causes Crohn's Disease. However, the major advances that have been made over the past few years have suggested that Crohn's is caused by a combination of factors:

- Genes – if you have a family member with Crohn's, you are more likely to get it
- A problem with the immune system can lead to an attack of the digestive system
- Smoking – leads and contributes to more severe disease and greater risk of having surgery

Managing Crohn's disease symptoms

- Even though diets don't cause Crohn's disease, paying attention to what you eat may help you reduce symptoms. Keeping a food diary may be beneficial for you to track how your diet relates to your symptoms
- Stop smoking – as well as raising the risk for developing Crohn's, smoking can also trigger flare ups
- Manage your stress levels as stress does strongly impact symptoms

Gastroesophageal reflux disease (GERD)

GERD occurs when stomach acid frequently flows back into the tube connecting your mouth and stomach (the esophagus) This acid reflux can irritate the lining your esophagus. GERD and heartburn are different but related. Heartburn is a symptom of acid reflux, and GERD is a condition characterized by severe acid. Many people experience mild GERD from time to time and most people can manage the discomfort it causes by lifestyle changes or over the counter medication. However, some people experience severe symptoms and need further medical help from their doctor.

GERD is a chronic disease. Treatment usually must be maintained on a long-term basis, even after symptoms have been brought under control. Concerns of daily living and compliance with long-term use of medication need to be addressed with your doctor as well.

GERD is often characterized by painful symptoms that can undermine an individual's quality of life. Various methods to effectively treat GERD range from lifestyle measures to the use of medication or surgical procedures. Among these symptoms are:

- A burning sensation in your chest (heartburn) usually after eating
- Chest pain
- Regurgitation of food or sour liquid
- Difficulty swallowing
- Nausea or vomiting
- Sore throat

- Bad breath
- Bloating and feeling sick

There are often no reasons why GERD occurs, but it can be made worse by:

- Certain foods and drinks like; tomatoes, fatty foods, spicy foods, alcohol, coffee
- Being overweight or obese
- Smoking
- Pregnancy – a change in hormones
- Stress
- Some medicines

The goals of treating GERD are to bring the symptoms under control so that the sufferer gets the needed relief, to heal the inflamed esophagus, to manage or prevent complications like strictures, and finally, to maintain the symptoms of GERD in remission so that daily life is slightly impacted by reflux.

Treatment options for GERD include lifestyle modifications, medications, and surgery. GERD is a recurrent and chronic disease for which long-term medical therapy is usually effective. It is important to recognize that chronic reflux does not resolve itself. There is not yet a cure for GERD. Long-term and appropriate treatment is necessary. Nonetheless, over-the-counter medications can only provide temporary symptom relief, but they do not prevent recurrence of symptoms or allow an injured esophagus to heal.

Therefore, over-the-counter medications should probably not be taken regularly as a substitute for prescription medicines as they may be hiding a more serious condition. If needed regularly, for more than two weeks, consult a doctor and get appropriate treatment.

Coeliac disease

Coeliac disease is an autoimmune condition which occurs in people who become sensitive to a protein called gluten in their diet. Gluten is found in wheat, barley and rye and is normally a nourishing and harmless part of the food we eat. However, if you have coeliac disease, gluten causes damage to the lining of your small intestine which can cause problems with the absorption of nutrients and vitamins that you eat. Coeliac disease can be diagnosed at any age, from infancy to old age, although most people diagnosed are over 40. The treatment, which is usually very successful, is to remove all sources of gluten from the diet. Coeliac disease is a life-long condition for which there is no cure at present.

Coeliac disease is hereditary, and it runs in families. People with a family history of coeliac disease relative with coeliac disease, are at a higher risk of developing coeliac disease.

Recent research has shown that approximately one in 100 people in the UK have this condition and it is known to occur more frequently in Caucasian populations in Europe and also, in developing countries where wheat is a staple diet (the west of Ireland has the highest rate of the disease in the world).

The incidence of coeliac disease in people with first-degree relatives (parent, child, sibling) who have coeliac disease is 1 in 10, and with second degree relatives (aunt, uncle, cousin) who have coeliac disease is 1 in 39.

Some people who have coeliac disease don't know they have it but may still have some mild symptoms and it is thought that only 1 in 800 people have been correctly diagnosed with coeliac disease in the UK.

Undiagnosed or untreated coeliac disease can lead to long term health conditions like:

- Iron deficiency anemia
- Early onset osteoporosis or osteopenia
- Infertility and miscarriage
- Lactose intolerance
- Vitamin and mineral deficiencies
- Central and peripheral nervous system disorders
- Pancreatic insufficiency
- Intestinal lymphomas and other GI cancers (malignancies)
- Gall bladder malfunction
- Neurological manifestations, including ataxia, epileptic seizures, dementia, migraine, neuropathy, and myopathy

Coeliac disease can develop at any age after people start eating foods or medicines that contain gluten. Left untreated, coeliac disease can lead to additional serious health problems. Among the range of symptoms of coeliac disease are:

- Diarrhea
- Abdominal pain or stomach ache
- Bloating and gas

- Indigestion
- Constipation
- Nausea and vomiting
- Weight loss

Other general symptoms are:

- Headaches
- Joint pain
- Itchy skin, skin rash
- Loss of bone density
- Anemia from iron deficiency
- Mouth ulcers
- Depression, stress, anxiety
- Nerve damage in the extremities called peripheral neuropathy which can cause tingling in the legs and feet
- Fatigue

Gluten can be found in three types of cereals:

1) Wheat
2) Barley
3) Rye

Gluten can also be found in any food that contains the above; Some of these foods are:

- Most baked products like cakes, cookies, pastries, doughnuts, pancakes, waffles, muffins
- Wheat based products like noodles, spaghetti, pasta, couscous, dumplings

- Some snack foods such as cereals biscuits, candy bars, granola bars, cookies, energy bars, pretzels
- Sauces – barbecue sauce, salad dressings, cream sauces, marinade, gravy mixes, malt vinegar, ketchup, soy sauce
- Certain beverages – beer, bottled wine coolers, drink mixes, coffee drinks, commercial chocolate milk
- Several processed foods – hotdogs, processed cheese, canned soups, puddings, instant desert mixes, certain ice creams, breakfast cereals, French fries, and other fried foods

Managing coeliac disease

Currently, the only treatment for coeliac disease is lifelong adherence to a strict gluten-free diet. People living gluten-free must avoid foods with wheat, rye and barley,

For most people with coeliac disease, switching to a gluten-free diet greatly improves the symptoms. And fortunately, there's a large section of gluten-free foods like; fish, meat, poultry, seafood, fruits and vegetables m, some grains such as rice, rice quinoa, rice flour, cereals like corn, millet, sorghum, diary, beans, legumes, nuts, cassava, soy, flax, gluten-free oats, potatoes, tapioca, corn, amaranth, arrowroot, nut floors.

Additionally, if you have been diagnosed with Coeliac disease, educate yourself about a gluten-free diet. The right diet will prevent damage of the lining of your intestines and the associated symptoms, such as diarrhea and stomach pain.

Diverticulitis and diverticulosis disease

Diverticulitis and diverticulosis are related digestive conditions that affect the large intestine. Together they are known as diverticular disease. Diverticula are small bulges or pockets that can develop in the lining of the intestine as you get older. Most people with diverticula do not get any symptoms and only know they have them after having a scan for a different reason. When there are no symptoms, the condition is called diverticulosis. When diverticula cause symptoms such as pain in the lower stomach, it's called diverticular disease. If the diverticula becomes inflamed or infected causing more severe symptoms, it's called diverticulitis.

Diverticulosis and diverticulitis are equally common in men and women, but it tends to affect older people more than the young. You are more likely to get diverticulosis disease if you don't get enough fibre in your diet. It is also thought that development of this disease may be a result of consuming over-refined foods.

Symptoms of diverticulosis and diverticulitis disease

- Stomach pain, usually in the lower left side
- Swelling and bloating
- Constipation or diarrhea or both
- Occasional blood in the stools
- Constant or severe stomach pains
- A high temperature – fever
- Having mucus or blood in the stools (rectal bleeding)

People most likely to get diverticulosis and diverticulitis disease

- Those over 50 years of age
- If you're overweight or obese
- Consuming a low fibre diet – if you eat less fruit and vegetables, beans, legumes, grains, nuts, and seeds
- Lack of exercise also contributes to this disease
- A diet high in fat and red meat
- Those who smoke
- Certain medications

Preventing and managing diverticulosis and diverticulitis disease Preventing constipation and straining while opening the bowels is crucial to preventing and reducing complications of diverticulosis and diverticulitis disease. This should be by:

- Eating more fibre in the diet as fibre makes the stools softer and bulkier and therefore easy to move about.
- Drinking plenty of water to keep the stool soft
- Daily exercise – helps the food to move easily through the digestive tract
- Stop smoking
- Lose any extra weight

Cancers of the digestive system

Like other parts of the body, the digestive tract can be a site for cancers, among these cancers are:

- esophageal or gullet cancer
- stomach or gastric cancer

- pancreatic cancer
- bowel cancer and liver cancer.

Esophageal cancer

Esophageal cancer is a cancer that occurs in the esophagus. The esophagus is a hollow muscular tube that's responsible for moving food from the throat to the stomach.

Esophageal cancer can occur when a malignant tumor forms in the lining of the esophagus. This cancer usually begins in the cells of the esophagus. As the tumor grows, it can affect the deep tissues and muscle of the esophagus. A tumor can appear anywhere along the length of the esophagus, including where the esophagus and the stomach meet. Men are more likely at risk of this cancer than women. Even so, other risk factors are:

- smoking
- being overweight
- excess alcohol consumption
- experiencing constant heartburn and;
- nutritional diets such as a diet high in smoked and pickled foods.

There are two common types of esophageal cancer:

1) Squamous cell carcinoma – cancer starts in the flat thin cells that make up the lining of the esophagus. This form of cancer most often appears in the top or middle of the esophagus, but it can appear anywhere.
2) Adenocarcinoma cancer occurs when cancer starts in the glandular cells of the esophagus that are responsible for the production of fluids such as

mucus. Adenocarcinomas are most common in the lower portion of the esophagus.

Symptoms of esophageal cancer:

- Difficulty swallowing
- Indigestion
- Pain in the throat area
- Vomiting
- Persistent cough
- Food coming back up the esophagus
- Chest pain
- Unexplainable weight loss
- Constant choking while eating
- Persistent hiccups
- Fatigue
- Hoarseness

Although it's not exactly clear what causes esophagus cancer, there's so much that can be done about lifestyle adjustments to prevent or reduce your risk of getting it.

Who's at risk of getting esophageal cancer?
Experts believe that the irritation of esophageal cells contributes to the development of cancer. Some habits and conditions that can trigger irritation include:

- Smoking
- Being overweight
- Excess alcohol consumption
- Having GERD
- Not eating enough fruit and vegetables in your diet

- Older males are at a higher risk of getting it than women
- Esophageal cancer happens more in African Americans and Asians

Lifestyle changes:

- Quit smoking
- Drink alcohol in moderation
- Maintain a healthy weight
- Eat more fruit and vegetables in your diet

Stomach cancer
Stomach cancer begins when cancer cells form in the inner lining of your stomach. These cells can grow into a tumor. Also called gastric cancer, the disease usually growsslowly over many years. Stomach cancer is most often seen in older people, especially in age groups of the 60s through to the 80s. Stomach cancers tend to develop slowly over many years. Before a true cancer develops, pre-cancerous changes often occur in the inner lining of the stomach. These early changes rarely cause symptoms, so they often go undetected.

Cancers starting in different sections of the stomach can cause different symptoms and tend to have different outcomes. The cancer's location can also affect treatment options. Stomach cancers are classified according to the type of tissue they start in:

- adenocarcinomas are the most common stock cancers; they start in the glandular

- stomach lining. There are 2 main types of stomach adenocarcinomas:
- The intestinal type tends to have a slightly better prognosis. Here, the cancer cells are more likely to have certain gene changes that might allow for treatment with targeted drug therapy.
- The diffuse type tends to grow and spread more quickly. It is less common than the intestinal type, and it tends to be harder to treat.
- Lymphomas develop from lymphocytes, a type of blood cell involved in the immune system. Lymphomas usually start in other parts of the body, but some can start in the wall of the stomach. The treatment and outlook for these cancers depend on the type of lymphoma and other factors.
- Sarcomas involve the connective tissue. Soft tissue sarcomas are a group of rare cancers affecting the tissues that connect, support, and surround other body structures and organs. Tissues that can be affected by soft tissue sarcomas include fat, muscle, blood vessels, deep skin tissues, tendons, and ligaments.

Scientists don't know exactly what makes cancer cells start growing in the stomach. But they do know a few things that can raise your risk for the disease. One of them is infection with a common bacteria H. pylorus, which causes ulcers. Inflammation in your gut called gastritis, a certain type of long-lasting anemia called pernicious anemia, and growths in your stomach called polyps also can make you more likely to get cancer.

Others are:

- Smoking increases the risk of getting this cancer. Smoking tobacco increases the risk of developing several types of cancer, including cancer of the stomach and the risk increases with the number of cigarettes smoked. 1 in 5 stomach cancers are reported to be related to smoking.
- Obesity – Having an unhealthy body weight: Having an unhealthy weight, particularly a body mass index (BMI) of over 30 kg/m carries an increased risk of developing stomach cancer. This can occur at the top part of the stomach (called the cardia) as well as cancers between the stomach and the gullet or the gastro-esophageal junction. 5 in 100 stomach cancers are reported to occur in people who have a high BMI.
- Drinking alcohol regularly has been reported to increase the risk of stomach cancer in some studies, and although the risk was felt to be conflicting in some people, it is still best to moderate alcohol intake and to stay within healthy guidelines as high alcohol intake can result in other damaging effects on health.
- A family history of stomach cancer: People who have close family members (brother, sister, or parent) with stomach cancer, have a higher-than-average risk of developing stomach cancer themselves. This is possibly due to eating a similar diet, having an infection with the H. pylori bacterium, or possibly sharing the same genetic background.
- Some people with a strong family and genetic risk

of stomach cancer, can develop it at a younger age-earlier than 50.

- Not consuming adequate amounts of fruit and vegetables or eating a diet high in salt can also increase the risk of developing stomach cancer. To emphasize, this cancer is also associated with a higher intake of red meat and particularly smoked meat products which are high in salt.
- Some types of previous stomach surgery may result in less stomach acid being produced, and this may also lead to an increased risk of stomach cancer.
- Long term stomach inflammation

The symptoms of stomach cancer

- Loss of appetite
- Weight loss
- Indigestion
- Heartburn or acid reflux
- Tiredness due to low red blood cells
- Pain at the top of the stomach
- Feeling sick
- Nausea
- Vomiting

Pancreatic cancer
Pancreatic cancer occurs in the pancreas, it is seldom detected and hard to diagnose. This is because the pancreas is buried deep within the body. Symptoms often appear when the cancer is in its advanced stages. For this reason alone, there aren't any early signs of pancreatic cancer.

Even when the cancer has advanced, most of the common symptoms can be subtle.

Pancreatic cancer is more common in older people. More than 45 out of 100 people diagnosed are aged 75 and over. This cancer is uncommon in people under 40 years old. Symptoms may include:

- Loss of appetite
- Jaundice or yellowing of the skin or whites of your eyes
- Low back pain
- Unexplained weight loss
- Blood clots
- Feeling tired/no energy
- Abdominal pain
- Bowel movement problems
- Fatigue
- Feeling shivery/ a high temperature

Doctors don't know what causes most pancreatic cancers. But there are some factors that may increase your risk of developing it. However, having any of these risk factors does not mean that you will develop cancer.

- Gender – more men are diagnosed with pancreatic cancer than women.
- Age – the risk of getting this cancer increases with age. Most people who develop pancreatic cancer are 45 years and above.
- Race/ethnicity – black people are more likely than white, Asian, or Hispanic to get pancreatic cancer.
- Family history – pancreatic cancer may run in the

family and may be linked with genetic conditions that increase the risk of other types of cancer.

- Diabetes – many studies have detailed that diabetes increases the risk of developing pancreatic cancer especially when a person has had diabetes for several years.
- Obesity – eating foods high in fat is a risk factor for pancreatic cancer. Research has shown that obese people have a high chance of being diagnosed and dying from pancreatic cancer.
- A diet high in fats leads to piling on the weight and puts you at a greater risk of the disease.
- Alcohol use can also push up your chances of getting pancreatic cancer and causing recurrent pancreatitis (the repeated inflammation of the pancreas)
- Smoking – smoking cigarettes, cigars, pipes and chewing tobacco all increase pancreatic cancer risk. The best way for people who smoke to reduce their risk of cancer and improve their overall health, is to stop smoking completely. The risk of pancreatic cancer in people who stopped smoking 20 years ago is the same as for people who have never smoked.
- Gallstones. People with gallstones have an increased risk of pancreatic cancer compared to those without gallstones. This may be because gallstones can cause chronic pancreatitis, which is another risk factor for pancreatic cancer.
- Women with metabolic syndrome have an increased risk of pancreatic cancer compared with those without the condition.

Pancreatitis

Pancreatitis is the inflammation of the pancreas. The two forms of pancreatitis are acute and chronic.

1) Acute pancreatitis is sudden inflammation that last a short period of time. Most people with a acute pancreatitis start to feel better within a week after getting the right treatment with no further complications. Severe cases of acute pancreatitis can cause tissue damage, bleeding, infection or even cause damage to other organs of the body.

2) Chronic pancreatitis is when the pancreas becomes permanently damaged from inflammation over many years. This can occur after lengthy episodes of acute pancreatitis or due to heavy alcohol consumption.

Symptoms of pancreatitis

- Sudden severe pain in the center of your abdomen
- Feeling sick
- Diarrhea
- A high temperature or fever
- Nausea or vomiting
- Tenderness when touching the abdomen
- Unexplained weight loss
- A rapid pulse
- Higher heart rate

Causes of pancreatitis

Acute pancreatitis is often linked to:

- Drinking too much alcohol
- Gallstones stuck in the bile duct preventing pancreatic juices from reaching the parts of the digestive tract they are needed
- Certain medications
- Obesity
- Pancreatic cancer
- Infection
- Abdominal surgery
- Trauma

Chronic pancreatitis causes:

- Cystitis fibrosis
- A family history of pancreatic disorders
- Gallstones
- Medications
- Long term alcohol consumption

Because many cases of pancreatitis are caused by alcohol abuse, managing this habit will require you to focus on limiting how much you drink or not drinking at all. If your drinking is a concern, talk to your doctor or health care professional about an alcohol treatment center. A support group such as Alcoholics Anonymous could also help.

Fatty liver disease
Fatty liver also known as hepatic steatosis. It happens when extra fat builds up in the liver. Heavy alcohol drinking means that you're more likely to get fatty liver disease as too much alcohol causes fat to build up inside your liver cells, making it difficult for the liver to work

efficiently. Too much fat in the liver can also cause liver inflammation. However, people who don't consume a lot of alcohol can also get fatty liver disease.

Symptoms of alcohol-related liver disease (ARLD)
ARLD does not usually cause any symptoms until the liver has been severely damaged. When this happens, symptoms can include:

- Feeling sick
- Weight loss
- Loss of appetite
- Yellowing of the eyes and skin (jaundice)
- Swelling in the ankles and tummy
- Confusion or drowsiness
- Vomiting blood or passing blood in your stools
- Chronic fatigue
- Nausea
- Bruising easily

Preventing ARLD
There's currently no specific medical treatment for ARLD. The main treatment is to stop drinking, preferably for the rest of your life. This will reduce the risk of further damage to your liver and gives it the best chance of recovering. When someone is dependent on or addicted to alcohol, quitting drinking can be very difficult. But support, advice and medical treatment may be available through local alcohol support services. A liver transplant may be required in severe cases where the liver has stopped functioning and does not improve when you stop drinking

alcohol. You will be considered for a liver transplant if you have developed complications of cirrhosis despite having stopped drinking. A liver transplant requires a person to completely give up alcohol while awaiting the transplant, and for the rest of their life.

The most effective way to prevent ARLD is to stop drinking alcohol or stick to the recommended limits: Men and women are advised not to regularly drink more than 14 units a week. Even if you have been a heavy alcohol user for many years, reducing or stopping your alcohol intake will have important and positive short – and long-term benefits for your liver and overall health.

Non-alcoholic fatty liver disease (NAFLD)
Non-alcoholic fatty liver disease is the term for a range of conditions caused by a build-up of fat in the liver. It's usually seen in people who are overweight or obese. Early stage NAFLD does not usually cause any harm, but it can lead to serious liver damage, including cirrhosis if it gets worse. Having high levels of fat in your liver is also associated with an increased risk of serious health problems, such as diabetes, high blood pressure and kidney disease. If you have diabetes, NAFLD increases your chance of developing heart problems. If detected and managed at an early stage, it's possible to stop NAFLD getting worse and reduce the amount of fat in your liver. You're at an increased risk of NAFLD if you:

- Are obese or overweight-particularly if you have a lot of fat around your waist
- Have type 2 diabetes

- Have high blood pressure
- Have high cholesterol
- Have metabolic syndrome (a combination of diabetes, high blood pressure and obesity)
- Are over the age of 50
- Smoke

Symptoms of NAFLD

There are not usually any symptoms of NAFLD in the early stages. You probably will not know you have it unless it's diagnosed during tests carried out for another reason.

Occasionally, people in the advanced stages of NAFLD may experience:

- Pain in the top right of the tummy (over the lower right side of the ribs)
- Extreme tiredness
- Unexplained weight loss

Gallstones

Gallstones are small stones made of cholesterol that form in the gallbladder. Gallstones range in size from as small a grain of sand to as a large as a golf ball. Some people develop just one, while others develop many stones at the same time. There are two main types of gallstones: (a) Cholesterol gallstones are usually yellow green in color and are made of mostly hardened cholesterol, they may form if there is too much cholesterol in the bile and are the most common type. (b) Pigment gallstones are dark in color and are made of bilirubin. They form when the bile has too much bilirubin. Pigment gallstones are more common among people with liver disease, infected bile tubes, or blood disorders, such as sickle-cell anemia. Gallstones are very common and do not cause any symptoms. People who experience symptoms from their gallstones often require surgery – the removal of the gallbladder. Gallstones that don't cause any signs and symptoms usually don't require treatment.

Many people with gallstones experience no symptoms at all. This is because the stones stay in the gallbladder and cause no problems. Gallstones are very common and some studies indicate that one in six men and one in three women suffer from gallstones at some point in their life. The prevalence of gallstones in Europe is around 10–15%.

By the age of 60 nearly a quarter of women (and a rather smaller number of men) will have developed some gallstones. The condition is much more common in women especially those who have had children and who are overweight. However, recent research shows that the incidence of gallstones in much younger women and even teenagers have been rising. Overall gallstones seem to be more common generally, possibly because of changes in our diet over the last two generations.

The primary symptom of gallstones is pain that comes on suddenly and quickly gets worse. This pain can occur on the right side of the body, just below the ribs, between the shoulder blades, or in the right shoulder.

Other symptoms include:

– Nausea and vomiting
– Yellowing of the skin and whites of the eyes (jaundice)

- Chest pain
- Fever and chills
- A high temperature
- Sweating
- Restlessness

Gallstones may form when the chemicals in the gall-bladder are out of balance, such as cholesterol, calcium bilirubinate, and calcium carbonate.

Experts are not sure why gallstones occur in some people and not in others. But it is thought that they may be a result of:

- Too much cholesterol in the bile. This happens when the liver excretes more cholesterol than your bile can dissolve, then the excess cholesterol may form into crystals and eventually into stones.
- Too much bilirubin in the bile, bilirubin is the chemical that's produced when your body breaks down red blood cells. Excess bilirubin contributes to gallstones formation.
- When the bladder doesn't empty correctly and bile becomes very concentrated, contributing to a formation of gallstones.

Risk factors of gallstones include:

- Being over 60 are more at risk
- A family history of gallstones
- Having been pregnant
- Having diabetes type 2
- Having Native American or Mexican heritage

- Taking medication with a high estrogen content of certain birth controls
- Taking drugs that lower cholesterol
- Living with cirrhosis
- Losing weight rapidly
- Eating a low fibre diet
- Eating a diet high in fat and cholesterol
- Being overweight or obese
- Gender – more women develop gallstones than men

Hemorrhoids (Piles)

Hemorrhoids or piles are enlarged blood vessels in and around the lower rectum and anus. Piles usually look like small round discolored lumps. You may be able to feel them at the opening of your bottom. They often get better on their own after a few days. Although in extreme cases with excessive bleeding and severe pain you should seek medical help.

Causes of hemorrhoids

Piles develop when the veins in the anal canal become swollen, which may happen for a number of reasons:

- If you strain when you are opening your bowels, this may be because you are constipated or with long lasting diarrhea
- Getting older weakens the anal canal which makes piles more likely
- Having a consistent cough
- Lifting heavy objects
- Pregnancy may also cause piles due to the higher

pressure in the tummy area when you are pregnant
- Obesity
- Eating a low fibre diet

Symptoms of hemorrhoids

- Blood in the stools – a bright red color
- A lump around the anus
- Anus itchiness
- Pain and discomfort during and after you open your bowels
- Mucus discharge from the anus
- Feeling like you still need to pass stools soon after opening your bowels

Preventing and managing hemorrhoids

The best way to prevent hemorrhoids is to keep your stools soft so that they can be passed easily without straining. Other tips are:

- Eating high fibre foods
- Drinking plenty of fluids
- Exercise regularly
- Avoid straining
- Open your bowels as soon as you feel the urge to do so, putting off passing stools dry them and hardens them, therefore making them difficult to pass out.
- Avoid sitting, particularly on the toilet, for lengthy periods as this increases the pressure on your veins in the anus.

Lactose intolerance

Lactose intolerance is a common digestive problem where the body is unable to digest Lactose, a type of sugar mainly found in milk and dairy products. Most adults cannot process milk, this is because infants produce the enzyme lactase which is required to digest the milk sugar lactose in mother's milk. As a child gets older, the ability of the body to produce lactose decreases. That's why very few adults have the capacity to properly digest milk and milk products. All human beings have the genes to digest lactose, though in exceedingly scarce cases problems with lactose intolerance can occur from birth. Newborns born with this rare condition are unable to digest their mother's breast milk, and drinking it causes severe diarrhea.

If you are lactose intolerant, your body struggles to absorb the lactose from food into your body through the small intestines. The lactose instead goes into your colon, where it ferments with normal bacteria, causing diarrhea, irritation, bloating, and gas.

Many people confuse lactose intolerance for having a dairy allergy when it's not the case. if you have an allergy to dairy, you'll be allergic to certain proteins in dairy products. Usually, the symptoms of lactose intolerance are less severe than those of a dairy allergy. The symptoms of lactose intolerance are very uncomfortable, but not dangerous, most people can manage their symptoms by changing their diet and eliminating the amount of lactose they consume. Other people cut the lactose out of their diets completely.

Symptoms of lactose intolerance

- Stomach pain and cramps
- Diarrhea
- Bloating
- Stomach rumbling
- Feeling sick
- A windy stomach

You can manage your lactose intolerance by:

- Cutting out lactose or eliminating it from your diet
- Consume only lactose free products such as soy milk, lactose free cow's milk, rice milk almond milk, hazelnut milk, coconut milk and potato milks

Allergies

An allergy is the response of the body's immune system to normally harmless substances like foods, pollen, mold, insects, latex, medicines, pets. Whilst in most people these substances pause no problems, in allergic people, their immune system identifies them as threats and products a negative reaction. Allergies are one of the most common chronic diseases that it can last for a long time. What causes allergy is yet to be determined, but research continues into what lies behind the increase in this disease across the in all age groups. Allergies are most common in children, but they can occur at any age, and they can run in families – if you have an allergy, it is likely that one of your parents had allergy too.

Food allergies

Many people mistake food allergies for food intolerance, even though both conditions have similar symptoms, they are not the same thing. However, both conditions usually irritate the digestive tract with uncomfortable symptoms such as:

- An itchy rash
- Swelling of the lips, tongue, face and throat and other parts of the body
- Shortness of breath
- Abdominal pain
- Nasal congestion
- Dizziness
- Diarrhea
- Nausea
- Vomiting

Types of food allergies

1) IgE-mediated food allergy is the most common type. It's triggered by the immune system producing an anybody called immunoglobulin E(IgE) symptoms can occur within seconds or minutes of after eating. With this type of allergy, the risk of anaphylaxis is high.

2) Non IgE-mediated food allergies are not caused by immunoglobulin E unlike the previous type. This type of allergy is usually difficult to diagnose as symptoms take a while to present themselves, they may take up to several hours.

3) Mixed IgE and non IgE mediated food allergies. In such cases, symptoms from both types can be experienced by some people.

The most extreme cases of food allergy can cause an extreme allergic reaction called anaphylaxis. This can cause life threatening symptoms to appear very quickly, like:

- A dramatic fall in blood pressure
- Rapid pulse
- Dizziness or loss of consciousness
- A swollen throat that makes it difficult to breathe
- Constriction and tightening of the airways
- Shortness of breath

In adults, most food allergies are triggered by certain foods, these include:

- Peanuts
- Tree nuts like walnuts, almonds, Brazil nuts, pecans, pine nuts
- Eggs
- Fish
- Shellfish like crab and lobster
- Some fruit and vegetables
- Soy
- Wheat

Less common trigger foods are:

- Seeds
- Corn
- Meat

- Spices
- Gelatin

Managing allergies

There's currently no cure for food allergies, although many children will grow out of certain ones like allergies to milk and eggs. The most effective way to prevent symptoms is to eliminate the offending food-known as an allergen, from the diet. However, with children, it's crucial to check with the doctor first before eliminating certain foods.

Stomach ulcers

Stomach ulcers also known as gastric ulcers are painful sores in the stomach. Stomach ulcers occur when the thick layer of the mucus that protects the stomach digestive juices is reduced. This allows the digestive acids to eat away at the tissues that line the stomach causing an ulcer. Stomach ulcers may be easily treated, but can become severe without proper treatment, or can come back after treatment of the underlying cause is not addressed. Stomach ulcer symptoms depend on the severity of the ulcer. But the most common ones are:

- Passing dark sticky stools
- A sudden sharp pain in the tummy
- Bloating
- Nausea and vomiting
- Indigestion
- Heartburn
- Loss of appetite
- Weight loss

Severe complications are rare but can be very serious and life threatening, these are:

- Bleeding of the ulcer
- Blocking off food movement through the digestive tract

Stomach ulcers may also be caused by lifestyle factors, these can increase the chances of developing as well as making treatment for this disease less effective. These are:

- An infection with Helicobacter pylori (H. pylori) bacteria
- Taking anti-inflammatory medicines such as ibuprofen or aspirin for a long time or at high doses
- Stress
- Smoking
- Excess alcohol consumption

Food poisoning
Food poisoning is caused by bacteria, parasites, and viruses. But by far the most cause is bacteria. Dangerous bacteria like E. coli, salmonella and listeria are by far the biggest culprits of food poisoning cases. Most people have had an episode of food poisoning, especially when traveling abroad. Contaminated food may taste normal, with the symptoms appearing hours or days later. In most people, symptoms are mild and clear up in a few days. However, some severe cases will require treatment.

The salmonella bacteria
Salmonella is a group of bacteria that can cause gastro-

intestinal illness and fever called salmonellosis. Salmonella can be spread by food handlers who don't wash their hands, surfaces and tools used in food preparation and eating raw or uncooked foods.

Salmonella can get into our food in the most unexpected ways, but the most routes involve animals to humans especially with people who have direct contact with certain animals like poultry and reptiles. People can spread the bacteria from the animals to food if they don't wash their hands properly before handling food. Pets can also spread the bacteria within the home environment if they come into contact with food contaminated with salmonella.

When the salmonella bacteria are ingested, they pass through the stomach and populate the large and small intestines. There, they invade the intestinal mucosa and enter the bloodstream, causing infections in the blood, skin, bones, ears, eyes, urinary tract, heart, valves and lungs.

Symptoms of food poisoning
- Diarrhea
- Nausea
- Vomiting
- Stomach pain cramps
- A fever or high temperature
- Feeling generally unwell
- Loss of appetite
- Headaches

Food poisoning is especially serious and potentially life-threatening to pregnant women and their unborn babies, infants and young children, people with chronic illnesses and the elderly. These individuals should take extra precautions by avoiding the following foods:

- Raw or rare meat and poultry
- Raw or undercooked fish or shellfish, including oysters, clams, mussels, and scallops
- Raw or undercooked eggs or foods that may contain them, such as cookie dough and homemade ice cream
- Rice is likely to cause food poisoning due to a bacterium that is resistant to heat and so lives on after cooking
- Unpasteurized juices and ciders
- Unpasteurized milk and milk products
- Soft cheeses, such as feta, Brie, and Camembert; blue-veined cheese; and unpasteurized cheese
- Refrigerated pates and meat spreads
- Uncooked hot dogs, luncheon meats and deli meats
- Leafy vegetables and greens may also get contaminated especially when eaten raw for example brussels sprouts

Prevention of food poisoning

- Wash your hands thoroughly with soap and water and dry them before handling food, after handling raw food and touching the bin, using the toilet, blowing your nose, touching a pet etc.
- Wash cutting boards and knives with antibacterial soap and warm to hot water after handling raw meat, poultry, seafood, or eggs.

- Keep raw foods separate from ready-to-eat foods. When shopping, preparing food or storing food, keep raw meat, poultry, fish and shellfish away from other foods. This prevents cross-contamination. It's especially important to keep raw meat away from ready-to-eat foods, such as salad, fruit and bread. This is because these foods will not be cooked before you eat them, so any bacteria that get onto the foods from the raw meat will not be killed.
- Cook food properly and thoroughly. Make sure poultry, pork, burgers, sausages and kebabs are cooked until steaming hot, with no pink meat inside. Do not wash raw meat (including chicken and turkey) before cooking, as this can spread bacteria around your kitchen.
- Keep your fridge temperature below 5C and use a fridge thermometer to check it. This prevents harmful germs from growing and multiplying. To avoid overfilling your fridge and if it's too packed, air cannot circulate properly, which can affect the overall temperature.
- Refrigerate or freeze perishable foods promptly within a few hours of purchasing or preparing them.
- Defrost food safely.

To recap, when the digestive system is severely affected by certain diseases, surgery may be used as a treatment. This is true cases of cancer and in severe cases of inflammatory bowel disease (IBD). Some parts of the digestive tract can be removed in part or in full:

- The large intestine can be removed partially or fully in an ileostomy or colostomy. or J-pouch surgery. Most people live full and productive lives after these surgeries.
- The rectum and the anus can be removed, which is also called ileostomy or colostomy.
- Parts of the small intestine can be removed, but since this is where most nutrients are absorbed, an effort is made to keep it as intact as possible.
- Part of the stomach can be surgically removed, and people can live well after this surgery as well.

* * *

PART 5

*Enemies of the
Digestive System*

For a healthy digestive system, there are many habits you ought to avoid. If you're aiming at preventing digestive complications, changing your lifestyle can turn your life around for the better. And not only for this, but your entire health in general.

Alcohol

Research suggests that chronic alcohol consumption is associated with an increased risk of major gastrointestinal cancers including cancer of the esophagus, stomach, and colon (Colorectal cancer). The risk generally increases as alcohol consumption increases and in combination with other lifestyle-related factors, such as smoking tobacco or metabolic syndrome. And although alcohol was initially thought to act as a direct carcinogen, research instead suggests that alcohol-induced gut inflammation may be at fault.

Too much alcohol consumption can damage the digestive system and indeed your entire health. Once the alcohol enters the blood stream, it is not treated like other nutrients in food. What happens is, the digestive system works extra hard to eliminate it from the body.

Over-drinking can damage organs and parts of the GI tract, from the mouth, throat, esophagus, stomach, intestines, but mostly the liver. It's in the liver that alcohol is broken down into acetaldehyde, which poisons cells causing inflammation changes in the liver which can consequently lead to liver disease. Alcohol can also cause cell and tissue damage from the toxic by-products produced when alcohol is metabolized.

Effects of excess alcohol consumption on the digestive system

The mouth and throat: When you drink alcohol, it quickly penetrates the saliva in your mouth, and when converted to acetaldehyde, can damage the tissues in your mouth. A British study found that approximately one-third of mouth and throat cancers were caused by drinking alcohol.

The esophagus: once swallowed, alcohol can damage the cells of your esophagus and can increase your risk of cancer of the esophagus. It can also cause acid reflux.

The stomach: it's in the stomach that alcohol spends the most time. Alcohol can interfere with stomach function, affecting acid production, diminishing your stomach's ability to destroy harmful bacteria that enters the stomach and allowing it to enter your upper small intestine. This can also damage the mucous cells meant to protect your stomach wall from being damaged by acid and digestive enzymes, thereby causing inflammation. Strong alcoholic drinks-beverages with more than 15 percent, can delay stomach emptying, resulting in bacterial degradation of food and abdominal discomfort.

The intestines: All the food that is undigested passes from the small intestine to the large intestine and is then expelled from the body through the anus. Alcohol reaches the large intestine by way of your bloodstream, where it can increase the risk of bowel cancer. Moreover, drinking alcohol is associated with acid rising up from the stomach into the throat also known as acid reflux. Some evidence suggests alcohol can make the stomach produce more acid than usual which can gradually wear the stomach lining and make it inflamed and painful.

The liver: The liver is very resilient and capable of regenerating itself. However, each time your liver filters alcohol, some of the liver cells die. The liver can develop new cells, but prolonged alcohol misuse over many years can reduce its ability to regenerate. This can result in serious and permanent damage to your liver.

A healthy liver should contain very little or no fat, though after years of heavy drinking, it may develop alcoholic hepatitis where it becomes swollen. Complete failure of the liver is the end-stage or cirrhosis of the liver. This is liver damage where healthy cells are replaced by scar tissue and the liver is unable to perform its vital functions of metabolism, production of proteins, blood clotting factors and filtering toxins. Liver cirrhosis is a serious condition because once the liver becomes cirrhotic, the damage is irreversible-causing liver failure and eventual death.

Smoking
Smoking can harm the digestive system in many ways. This is because when you inhale the toxic smoke from the cigarettes, it goes into all parts of the body including the vital organs in the digestive system. The fact that the toxic smoke must pass through the digestive tract where food is converted into essential nutrients for survival

has serious consequences. Cigarettes are poisonous as they contain more than 4000 chemical compounds and hundreds of toxins.

What's in a cigarette?

- Nicotine: Nicotine is a very addictive drug. Most people who smoke regularly don't do so out of choice – they do it because they have a nicotine addiction. Some people smoke to relieve their stress. Compared to other substances in tobacco smoke, nicotine is the less harmless.
- Tar: Tar is a sticky-brown viscous substance that collects in the lungs when you breath in cigarette smoke. Tar contains cancer-causing chemicals. But it can cause more than just lung cancer. It also increases the risk of other lung diseases. This includes emphysema and chronic obstructive pulmonary disease (COPD).
- Carbon monoxide: Cigarette smoke contains a poisonous gas called carbon monoxide which can kill when inhaled in large amounts. Carbon monoxide stops your blood from carrying as much oxygen. This means your heart must work harder, and your organs don't get the amount of oxygen they need. This increases your risk of heart disease and stroke.
- Arsenic: a deadly poison
- Acetone: a chemical used in nail varnish
- Ammonia: Found in cleaning products
- Benzene: a chemical used as a solvent in fuel
- Ethanol: a chemical used in antifreeze

- Polonium: a highly radioactive element
- Formaldehyde: used as a preservative in laboratories and mortuaries
- Hydrogen cyanide was used in the gas chambers in World War II and is currently used in rat poison
- Butane and methanol are found in fuel
- Cadmium is a metal used in car batteries
- Phenol is used in fertilizers
- Naphthalene is a carcinogen used in moth balls

Given the large quantity of poison packed in cigarettes, one can easily see the tremendous damage that smoking does to the digestive system and the entire body. As a result, smoking contributes greatly to a host of illnesses such as GERD, Crohn's disease, pancreatic cancer, bowel cancer, liver cancer, stomach cancer, heartburn, IBS, pancreatitis, stomach ulcers, diverticular disease, and esophageal cancer.

Effects of smoking on the digestive system

The mouth and throat: mouth cancer and larynx cancer can appear anywhere in the mouth, including the inside of the lips, cheeks, gums, and larynx in the throat. Tobacco is carcinogenic, hence, it contains chemicals that can damage the DNA in cells and cause cancer. Smoking increases the risk of both mouth cancer and larynx cancer. According to the Mouth Cancer Foundation, approximately 90% of people with oral cancer are tobacco users, and smokers are six times more likely than non-smokers to develop oral cancer. Additionally, users of smokeless tobacco, commonly known as chewing tobacco, have a 50 times more likely chance of developing mouth cancer.

The esophagus: Using any form of tobacco, such as cigarettes, cigars, pipes, chewing tobacco, and snuff raises the risk of esophageal cancer.

The liver: The liver normally filters alcohol and other toxins out of your blood. But smoking limits your liver's ability to remove these toxins from your body. If the liver isn't working as it should, it may not be able to process medications well. Studies have shown that when smoking is combined with drinking too much alcohol, it makes liver disease worse.

The small and large intestines: An estimated 7% of bowel cancer cases in the UK are linked to tobacco smoking. Bowel cancer risk increases with the number of cigarettes smoked per day. Smokers are more likely to develop polyps (non-cancerous growths in the bowel) which could turn into cancer if not discovered.

The pancreas: Smoking is one of the most important risk factors for pancreatic cancer. The risk of getting pancreatic cancer is approximately twice as high among people who smoke compared to those who have never smoked. About 25% of pancreatic cancers are thought to be caused by cigarette smoking. Cigar smoking and the use of smokeless tobacco products also increase the risk. However, the risk of pancreatic cancer starts to drop once a person stops smoking.

Points to remember

- Smoking has been found to increase the risk of cancers of the mouth, esophagus, stomach, and pancreas. Research suggests that smoking may also increase the risk of cancers of the liver, colon, and rectum
- Smoking increases the risk of heartburn and gastro-esophageal reflux disease (GERD)
- Smoking increases the risk of peptic ulcers
- Smoking may worsen some liver diseases, including primary biliary cirrhosis and nonalcoholic fatty liver disease (NAFLD)
- Current and former smokers have a higher risk of developing Crohn's disease than people who have never smoked
- People who smoke are more likely to develop colon polyps
- Smoking increases the risk of developing pancreatitis
- Some studies have shown that smoking may increase the risk of developing gallstones

A poor diet

We are what we eat or not eat. That's a fact. Your nutrition can impact your digestive system in a positive or negative way. Your digestive system spends a lot of time processing food to produce energy, for growth, repair and getting rid of waste. How efficiently it does this depends on the quality of food you provide it with. Eating the wrong foods can lead to faulty digestion, poor absorption, inflammation, infection and illness. Besides, poor nutrition may increase the risk of developing certain long-term diseases such as diabetes, heart disease and strokes. Conversely, a poor diet may contribute to developing certain cancers, weight gain and can lead to increased risk of developing certain long-term diseases. Any of these health conditions can lead to a poor quality of life and other health complications, which can eventually result in a decreased life expectancy.

Intestines are an important part of the digestive system. On top of facilitating absorption and digestion of foods, they also help in building immunity. The intestinal microbiome is a complex system consisting of an uncountable number of microorganisms and bacteria that help in absorption of nutrients and generate energy, while protecting us from viruses and harmful bacteria. However, a poor diet and unhealthy lifestyles can negatively impact the functioning of the intestinal system.

Additionally, not getting enough fibre in your diet will lead to fibre deficiency. Fibre helps support your gut and microbiome health, so, lack of it may result into experiencing irregular bowel movement, constipation, blood sugar fluctuations, low energy levels, bloating a rise in cholesterol levels. There are a number of microorganisms that reside in our gut that are integral to maintaining a healthy immune system, and fibre is what feeds these microorganisms and allows them to do their job.

Without proper amounts of fibre, the health of your immune system may also be compromised.

Other unhealthy eating habits that may affect gut health include:

- Overeating – it's important to understand that digestive enzymes are only available in limited quantities, so the larger the meal you eat, the longer it takes to digest it. If you over-eat frequently for a lengthy period, food will remain in your stomach for a longer period of time and be more likely to turn into fat, which may lead to obesity a condition that has consistently shown to increase the risk of heart disease, strokes, type 2 diabetes, high blood pressure, colon cancer, gallbladder disease.

- Eating too fast – when you eat too fast, you swallow air, which can cause bloating of the stomach and wind which can all be upsetting and uncomfortable to put up with.

- Eating too close to bedtime – this increases the uncomfortable feeling of heartburn through the night as the food in your stomach gets pushed back up your esophagus more easily when you're laying down.

- Eating while slouching – a slouched posture after eating a meal can trigger heartburn caused acid reflux. Sitting in this position puts pressure on the stomach and forcing acid the wrong way. The pressure that continuous sitting in a poor posture can do to the lower intestines is significant too. Food moves through much slower and it's more difficult for the body to move our food around as intended.

Lack of exercise:
Physical inactivity can lead to a whole range of digestive disorders like constipation, bloating and wind. Regular

exercise or movement activates the gut while also increasing
intestinal activity. People who are less active experience constipation as the large intestine responds to activity. Good muscle tone is also important for regular bowel movements.

Medication
Medicines taken orally can affect the digestive tract in several different ways, whether prescribed, non-prescribed or over the counter drugs. While these medicines are usually safe and effective, they may present side effects in some people. A vast number of these side effects affect the digestive system as they travel through. For example:

- Diarrhea
- Heartburn
- Nausea
- Vomiting
- Constipation
- Bloating
- Pain in the abdomen
- Pain while swallowing or tablets getting stuck in the throat
- Over-the-counter laxatives

(Drugs for loosening stools) can cause permanent damage to the digestive system if used for a long time.

Several other types of medications can cause delayed gastric emptying. Delayed gastric emptying means that the muscles in the stomach that are responsible for emptying are slowed down, such that the food isn't moved out of the stomach at the rate that it should. For people who are diagnosed with gastroparesis-a disorder that causes the stomach to delay emptying, drugs that increase this slowdown effect can cause significant problems.

Some people are more at risk of developing stomach irritation after taking medicines for pain and inflammation relief and bringing down a high temperature, this includes older people or those who already have a history of stomach problems. Older individuals who take this medication on a regular basis are at risk for stomach irritation. A history of peptic ulcers or gastritis is also associated with a greater risk of complications after taking this medication.

Drugs like heroin, cocaine and others in this category can cause abdominal pain, bowel tissue decay, stomach, and intestinal ulcers. Sadly, when it comes to substance abuse, not everything may be reversible even after quitting. This is because there's enormous amounts of poison in these drugs and therefore potentially extremely serious and life threatening.

While heroin, cocaine or other drugs may provide you with a fast hit of pleasure, be mindful that the pleasure is only temporary but, the consequences on your health will be permanent.

Short term and long-term effects of drug use on the digestive system include:

- Decreased appetite
- Stomach pain
- Nausea

- Vomiting
- Blood in the stools

It's very common for many addicts to ignore the above problems, and instead of seeking help from their doctor or going to a drug rehab center to quit, they chose to continue with their drug use habit because of the addiction. And after a lengthy period of drug abuse, the symptoms worsen and develop into serious health complications. Such as:

- Decreased appetite leading to malnourishment and extreme eight loss
- Changes to metabolism
- Stomach ulcers
- Abdominal bleeding
- Reduced blood flow to the gastrointestinal system
- Perforation of the intestines
- Perforation of the small blood vessel in the abdomen
- Bowel tissue decay

Not drinking enough
Water is involved in literally every step of the digestive process, which is just another reason why staying adequately hydrated is so critically important to your health. Water is essential for good health. In digestion, water helps break down the food you eat, allowing the nutrients to be absorbed by your body. Water also aids in flushing waste from the intestine and helps lubricate the colon and moist the stools. If you don't get enough fluids in your body, you may experience constipation along with abdominal discomfort.

Not drinking enough will also lead to a dry mouth and throat with dry mucous membranes with no saliva, making it difficult to swallow and even breathe. Taking in an adequate amount of fluid can help to regulate your bowel movements, prevent constipation, and break down foods in combination with stomach acids and enzymes. Your body will use the available fluids to help food move through your digestive tract, and if there is not enough available, the result can be constipation or bloating with slow digestion. Stomach ulcers or acid reflux issues can also present themselves in more severe or prolonged periods of dehydration. Furthermore, it can be quite challenging for people who suffer from coeliac disease or other illnesses where they lose a lot of water through vomiting and diarrhea. It is particularly important for all those who suffer from dehydrating illnesses to monitor their fluid intake to help manage their symptoms.

Stress
Although stress is a mental state, it can physically affect our gastrointestinal system and the bacterial residents within it. A recent study found that high levels of stress can affect gut bacteria to a similar degree as a high-fat diet. Meaning that stress can alter gut bacteria, and gut bacteria can influence stress levels. Therefore, stress is another huge enemy of your digestive system, as it slows down digestion. The same goes for anxiety, worry, fear, nervousness. Even the good emotions like excitement can impact the digestion processes. We have all experienced time when we were stressed, nervous or excited about an

important event in our lives-be it a wedding, giving an important speech, meeting people for the first time, getting ready for an exam or driving test and so forth.

You may have noticed that during these episodes you felt sick-with butterflies in your stomach causing diarrhea and frequent trips to the toilet. In more severe cases, stress can cause a decrease in blood flow and oxygen to the stomach, which could lead to cramping, inflammation and escalate symptoms of digestive illnesses like indigestion, heartburn, GERD, IBS, and stomach ulcers.

The intestines have a tight barrier to protect the body from food related bacteria, however, stress can make the intestinal barrier weaker and allow gut bacteria to enter the body. Although most of these bacteria are easily taken care of by the immune system and do not make us sick, the constant low need for inflammatory action can lead to chronic mild symptoms.

When stressed, individuals may eat much more or much less than usual, but understanding which foods are beneficial or harmful to gut health under stress can be a challenge for many. It's common for people under stress to choose pleasurable and palatable foods irrespective of the caloric intake. In fact, the foods eaten during times of stress typically favor high fat and/or high sugar content. Stress also promotes alcohol overuse, smoking and drug use. It's not uncommon for some people to unwind a difficult and stressful day by drinking alcohol or puffing away to relax, and while there is nothing wrong with the occasional drink, overuse can negatively impact your digestive system and overall health.

Ignoring the urge to go

The digestive system like other systems of the body has innate intelligence in the way it ingests, digests, absorbs, and eliminate eliminates food. The body has a system for sending signals when the rectum is full and needs to be emptied. When the rectum is full, you will feel the urge to open your bowels, and unless there is a good reason for putting off going-it's essential to go right away. It's possible to put this urge off and wait until the right moment comes, and doing this occasionally is fine, but when you let 'Putting it off' become a habit for lengthy periods even days, it will affect your body. The longer the stools remain in the rectum, the more water is absorbed from it, hence hardening it and making it difficult to pass out.

How does the defecation reflex work?

When you eat, food moves from the mouth to the esophagus to the stomach. The food then passes through the small intestine to the large intestine to the rectum. The rectum is the final portion of the large intestine that connects to the anus, or the opening where the body releases the stools.

The defecation reflex is triggered when the muscles in the colon contract to move stool toward the rectum. This is known as a "mass movement." When enough stools move to the rectum, the amount of stool causes the tissues in the rectum to stretch. Inside these tissues are special "stretch" receptors designed to signal the brain when they are stretched. The defecation reflex triggers the two main sphincters around the anal canal. The first is the internal anal sphincter, which is a muscle that can't be controlled voluntarily. The second is the external anal sphincter, which is skeletal muscle that you have some control over. The defecation reflex occurs when the internal anal sphincter relaxes and the external anal sphincter contracts.

The recto-anal inhibitory reflex (RAIR) is an involuntary internal anal sphincter relaxation in response to rectal distention. After the defecation reflex is triggered, you can either delay, ignore or pass out the stools. Delay occurs when a person doesn't go to the bathroom immediately. There are muscles in the anal sphincter that cause the stool to move backward slightly. This effect reduces the urge to go. If you choose to go, your brain activates voluntary and involuntary muscles to move stool forward and out of your body. There are two main defecation reflexes:

1) The myenteric defecation reflex is responsible for increasing peristalsis and propelling stool toward the rectum. This eventually signals the internal anal sphincter to relax and reduce sphincter constriction.
2) The parasympathetic defecation reflex. While the motions of moving stool are similar, a person can voluntarily control the parasympathetic defecation reflex, but they can't control the myenteric one.

It's possible that a person can have a myenteric defecation reflex without the parasympathetic reflex. When this occurs, the urge to go to the bathroom may not be as strong as when both reflexes are working. Passing out hard dry stools can be painful and can cause fissures that may take time to heal. Ignoring the defecation reflex can also result in constipation. The defecation function is an important one as it helps the body to get rid of waste products, therefore shouldn't be put off or ignored for a healthy digestive system.

In more severe situations, avoiding opening your bowels can lead to incontinence or cause fecal impaction (when a hard, dry mass of stool becomes stuck in the colon or rectum) or gastrointestinal perforation (a hole in the wall of the gastrointestinal tract) It can also cause distension (where the person loses sensation in the rectum called rectal hyposensitivity.

Not getting enough sleep
Sleep is a state characterized by changes in the level of consciousness, unresponsiveness to the surrounding environment, and inactivity of voluntary muscles.

Sleep restores people physically and psychologically and it's vital to health as it's during this time that the body repairs itself. When you don't get enough sleep and rest, you are highly likely to suffer from a variety of health ailments including those of the digestive system. Much like other systems, when you sleep your digestive system continues to work but slower. And because during this time you're not ingesting anything, tissues, muscles and cells get the chance to be repaired during sleep. But if you're sleep deprived, your digestive system won't have enough time to rebuild itself and repair.

On top of it all, if you have poor sleep patterns, your body won't be able to systematically produce melatonin a sleep hormone and prolactin which have been found to improve the good bacteria in the intestines and help digestion.

Recent studies have shown a strong association between sleep deprivation and gastrointestinal diseases and that there is unique interplay between certain gastrointestinal diseases and sleep. Sleep deprivation makes you more vulnerable to inflammation disorders such as IBD and IBS, Irritable Bowel Syndrome, also sometimes known as inflammatory disorders. Some forms of IBD are often caused by a problem with the immune system-by which the immune cells start to attack the intestinal tissue causing widespread inflammation. Unfortunately, sleep deprivation may exacerbate this problem.

In addition, unhealthy sleeping patterns will make you more susceptible to stress, as it will have a negative impact on your mood. Yet stress is an enemy of your digestive system and can exacerbate conditions such as IBS, GERD, and ulcers.

Therefore decision making even in the most health-conscious people can be impacted. It will affect the way you crave food, which most likely means you will reach for unhealthy snacks more than you usually do.

* * *

PART 6

Keeping your Digestive System Healthy

Maintaining good digestive health doesn't happen by accident. It requires work, smart lifestyle choices, and the occasional checkup and test.

Eat a balanced diet

A healthy diet is rich in fibre, whole grains, "good" or unsaturated fats and omega-3 fatty acids, fresh fruits, and vegetables. Eating fresh fruit and vegetables is a great way to get nutrients and natural enzymes that can aid overall digestive health and function. All the above dietary components turn down inflammation, which can damage tissue, joints, artery walls, and organs. Going easy on processed foods is another element of healthy eating. Besides. In addition, foods tend to slow down the digestive process which can lead to constipation.

Eat a high-fibre diet

One of the best ways to improve your digestive health is by maintaining a diet that is high in fibre and rich in fruits, vegetables, nuts, seeds, legumes, and whole grains. This keeps the normal process of digestion running smoothly, helping to prevent constipation and maintain a healthy weight. In addition, a high-fibre diet helps prevent or treat conditions such as diverticulosis, IBS, and hemorrhoids.

Drink plenty of fluids

Water is involved in almost every step of the digestive process. From the moment food enters the mouth, saliva helps to moisten the food making it easy to chew and swallow.

Saliva too is a vehicle for enzymes and gastric juices that play a great role in chemically breaking down food. Water is also needed in form of gastric mucus, which serves as a lubricant of food in order to facilitate movement within the stomach and the formation of a protective layer over the lining epithelium of the stomach cavity. This protective layer is a defense mechanism the stomach has against being digested by its own enzymes.

Water helps with the movement of waste along the entire length of the small and large intestines and is absorbed by the body through the colon wall. If you are not drinking enough water, the colon will get water from the waste material and pass it on to the body where it's needed for other important functions. However, drinking plenty of water will help to maintain body hydration and reduce the need for the colon to extract excess amounts of water from the waste material that's making its way to the colon. Waste will therefore maintain its softness and move through the large intestine with ease.

Limit or cut out alcohol
For some people, deciding to quit their alcohol drinking habit may be simple, but for others it may be a struggle due to dependence on alcohol for a long time such that the habit may have morphed into addiction. Deciding to quit drinking alcohol can come with several benefits. Whether you've been heavily relying on alcohol for a while or

mildly consuming it, anyone can experience the positive physical and mental changes.

Regular consumption of alcohol can damage the pancreas, which is a vital organ to digestion. Alcohol hinders vitamin and nutrient absorption in the small intestines and can cause chronic diarrhea and nausea in people who drink heavily. In addition, the transporting of toxins through intestinal walls is increased with alcohol consumption. The liver is responsible for breaking down alcohol and getting rid of nasty toxins.

Over time, alcohol use can cause the liver to become overloaded with toxins and a build-up of fat, which leads to steatosis, or "fatty liver," which is an early sign of liver disease. A fatty liver can lead to lead to hepatitis and cirrhosis. The good news is, some of these negative digestive effects can be reversed or improved when drinking is stopped.

Quit smoking

Quitting smoking can reverse some of the effects of smoking on the digestive system. For example, the balance between factors that harm and protect the stomach and duodenum lining returns to normal within a few hours of a person quitting smoking.

The effects of smoking on how the liver handles medicines also disappear when a person stops smoking. Quitting smoking can also improve the symptoms of other digestive diseases and keep them from getting worse. For instance, people with Crohn's disease who quit smoking have less severe symptoms than smokers with the disease. However, people who stop smoking continue to have a higher risk of some digestive diseases, such as colon polyps and pancreatitis, than people who have never smoked.

Keep up regular exercise

There is overwhelming evidence that keeping active has tremendous benefits to wellbeing. And it's also medically proven that people who do regular physical activities are less likely to develop; bowel cancer, type 2 diabetes, coronary heart disease, stroke, early death, hip fracture, constipation, depression, dementia, and many other illnesses.

For most people, the easiest way to get active is to take part in everyday life, like walking, cycling, running, gardening, doing house chores. However, the more you do, the better. Taking part in activities such as sports and exercise will have even more greater benefits for your health. Vigorous exercise or activity is also good for you as it raises heart rate, though you need to be moving quick enough to raise your heart rate. There is substantial evidence that vigorous activity can bring health benefits over and above that of moderate activity.

When you exercise, especially with high intensity exercises, your blood flow increases and pumps stronger through the intestinal muscles. This creates stronger contractions through the digestive tract to decrease transit time of food passing through the intestines. Muscular contraction is directly related to blood flow; less blood flow will mean weaker contractions, which equals slow food transit time. Exercises like running, jumping, dancing, skipping, yoga, swimming are especially good for relieving constipation.

Seek help to manage your stress

Your bowel movements can be affected by stress that you experience during the day. Everyone experiences stress at some point in life, however, severe stress can lead to problems with diarrhea and constipation as well as nausea and vomiting. Try finding ways to reduce your stress with stress relieving therapies like Cognitive behavioral therapy (CBT) is a technique that has been proven to help reduce anxiety and stress by helping you learn to replace negative, distorted thoughts with positive ones.

Cognitive Behavioral Therapy (CBT) is a family of talking therapies, all based on the idea that those thoughts, feelings, behaviors, and bodily sensations, are all connected. If we change one of these, we can alter all the others. When we're low or upset, we often fall into patterns of thinking and responding which can worsen how we feel.

CBT works to help us notice and change problematic thinking styles or behavior patterns so we can feel better. CBT is a collaborative therapy – it's not something that is done to someone, it's a way of working together with a CBT therapist on mutually agreed goals. CBT is mainly concerned with how we think and act now, although sometimes our current difficulties are related to things which have happened in our past, and so these might also be part of what we talk about. Goals for therapy are set together with the therapist after talking things through to properly understand the problem. Most sessions begin with agenda setting and agreeing together what that session will concentrate on. A therapist will not tell someone coming for therapy what to do or what to talk about.

Other relaxation techniques can also be of great stress relivers such as meditation, deep relaxation exercises, yoga, tai chi, or through an activity you enjoy like walking, reading or listening to music. Other factors important in reducing stress include getting enough restful sleep and eating a nutritious diet. Finding the best stress relief strategies may take some experimenting. Some strategies may work out for you, others may not. But it's important to get to the root of your stress trigger first before you initiate anything, as this will help you target your efforts at the source. Keeping your stress at a manageable level is important not only for your digestive health, but also for your overall well-being and therefore worth the effort.

Monitor your medication
Whilst digestive symptoms are usually harmless and often clear on their own, sometimes they may persist, this may be a sigh that something is serious. Certain prescriptions or over the counter medication can cause irritation, diarrhea, nausea, vomiting, constipation, abdominal pain. Certain medicines that your doctor may have prescribed for you for other health conditions can lead to side effects that may upset your tummy and cause indigestion, diarrhea, heartburn, irritation, or nausea, constipation, vomiting,

abdominal pain, bloating, ingestion, and other symptoms. Whether its prescribed or over the counter drugs, it's important to let your doctor know all the medications you're using so that it's easier for them to get to the root of your symptoms. For most people, NSAIDs (Non-Steroidal Anti-Inflammatory Drugs) and drugs for heartburn aren't going to cause major issues, but when stomach problems crop up, if these drugs are used on a regular basis, it could be a clue as to what might be causing the symptoms.

Some people have difficulty swallowing tablets or capsules, or sometimes take medicines without liquid. Tablets or capsules may get stuck in the esophagus, releasing chemicals that can irritate the lining of the esophagus. This may cause ulcers, bleeding, perforation, and narrowing of the esophagus. The risk of these types of injuries is greater in persons with medical conditions involving the esophagus,

Certain medicines can also cause ulcers in the esophagus when they become lodged there. These include aspirin and certain antibiotics. Some medicines interfere with the action of the sphincter muscle, located between the esophagus and stomach. Because this muscle is responsible for allowing the flow of food into the stomach after swallowing, any interference can increase the chances of reflux, or backup of the stomach's acidic contents into the esophagus.

One of the most common irritants to the lining of the stomach is that caused by NSAIDs. This includes medicines, such as ibuprofen, naproxen, diclofenac, and other common pain relievers. These medicines weaken the ability of the lining to resist acid made in the stomach and can sometimes lead to inflammation of the stomach lining (gastritis), ulcers, bleeding, or perforation of the lining. There's a greater risk in older people of irritation from these medicines because they are more likely to take these pain relievers for chronic conditions. People with a history of peptic ulcers and gastritis are also at risk.

A variety of medicines can cause constipation. For example, opioids like codeine, morphine. This happens because these medicines affect the nerve and muscle activity in the colon (large intestine) resulting in the slow and difficult movement and removal of stools. On the flip side, diarrhea is most often caused by antibiotics, which affect bacteria normally present in the large intestine. The presence of a bacteria called Clostridium difficile can cause colitis, resulting in very fluid watery stools. The most common medicines that cause diarrhea are antibiotics, antiacids, antidepressants, proton pump inhibitors and chemotherapy medicines used to treat cancer.

Note; if you think a medicine is causing any digestive symptoms, contact the doctor who prescribed it. The dosage may need to be changed, or the medicine may need to be completely stopped. Also, always read labels for over the counter and prescription medicines, as they provide important information on safe and proper medication use.

Many people do not read labels, or they may have trouble finding the information they need. If you are one of those people don't worry too much, as you are not alone! Before you start on any medication just make sure that you comb through the drug label with your doctor or pharmacist-understanding the information on medication labels is crucial to avoid errors especially in people with chronic illnesses and the elderly.

Don't ignore the urge to go

Having bowel movement happens at a different frequency for everyone. For instance, some people have a bowel movement every day, others once a week, every two days, once a week, this is all normal. When abnormal changes happen, you'll notice because you only know how what's your normal and how your body operates and when things are not right. Some people have a bowel movement about three times a week, while others, only once a week. The trick is not to ignore the urge to go and open your bowels as your body tells you.

If you are experiencing severe constipation and abdominal pain, talk to your doctor. You may just need to add a laxative to your routine, and this will help you get to a healthier, regular bowel opening routine. Additionally, you may need to adjust your lifestyle to get into a healthier routine. Your doctor will also advise you on whether medical testing should be done to rule out diseases affecting bowel movements. Some conditions that affect the large intestine can be treated easily, while others can be life-threatening, so don't put off consulting your doctor.

Maintain your gut bacteria

There are trillions of healthy bacteria naturally present in your digestive tract. Collectively, they are known as your gut microbes, and they're tremendously important for overall health especially in the digestion process. Out of the trillions of bacteria that live in our bodies, the ones in the gut may have the biggest impact on our well-being. The roles microbes play in digestive health are:

- to produce vitamin B12 and vitamin K
- control the growth of harmful bacteria
- break down poisons in the large intestine,
- break down some substances in food that cannot be digested, such as fibre and some starches and sugars.

They also produce enzymes that digest carbohydrates in plant cell walls. Most of the nutritional value of plant material would be wasted without these bacteria.

The food you eat is the main fuel for your gut microbes. That is why it's vital to consciously monitor your diet. Filling your daily diet with a range of foods is an excellent way to boost your gut microbes and your health generally. A review of the research literature suggests that diet can modify your microbes, resulting in a profound impact on your overall health. Some of the food that you eat does not absorb inside your body, for instance fibre, instead it travels straight to your lower digestive tract where it feeds the trillions of bacteria. This part of the fibre that feeds these microbes is called prebiotic. The most common prebiotic strains are Lactobacillus and Bifidobacterium. The idea is that, after you've swallowed these live microorganisms, they make their way through the stomach and take up residence in your gut. There is some evidence that prebiotics may be helpful in preventing diarrhea associated with antibiotics and improving symptoms of IBS, but more needs to be learned.

While feeding our microbes is beneficial, it can be detrimental when microbes are starved of fibre, as hunger-stricken microbes may start to feed on the protective mucus lining of the gut, which could lead to bowel inflammation diseases. Prebiotics are available in dietary supplements and in certain foods such as:

Yoghurt: yoghurt is one of the best sources of prebiotics. It is made from milk that has been fermented by prebiotics, mainly lactic acid bacteria and Bifidobacterium. In addition, yoghurt may be suitable for people with lactose intolerance. This is because, the bacteria turn some of the lactose into lactic acid. However, not all yoghurts contain live prebiotics.

Oats: are a great source of an indigestible form of soluble fibre called beta-glucans. These fibres not only feed your gut bacteria but have also been linked to improved insulin sensitivity as well as lower levels of "bad" LDL cholesterol. While all oats contain beta-glucans, raw oats are also a great source of resistant starch, which will provide you with its additional anti-inflammatory benefit.

Whole grains: The less processed a food is, the more it will travel down to your lower gut to feed your beneficial microbes. Replacing refined grains (via white breads, white pasta, and processed cereals) for fibre-rich whole grains, such as wheat, rye, and barley, is the easiest way to eat more prebiotic foods in your diet. Studies have found that adding whole grains to your diet, even if it's just eating a cup of whole-wheat breakfast cereal, can, after three weeks increase levels of Bifidobacterium and Lactobacilli.

Lentils: such as split peas, beans, and chickpeas are a source of resistant starch. These pass through the small intestine intact and make way into the large intestine as food for your gut bugs. The microbes ferment them into a fatty acid called butyrate, which helps to turn off the genes that lead to inflammation and insulin resistance. A recent study in the Journal of Functional Foods found that when you eat resistant starch, your gut biome gets stronger.

Fruit and vegetables: in general, are a great source of slow-digesting fibre and therefore, a great source of food for your microbes. Bananas are especially good eaten slightly green, as most of the starch turns to sugar as they ripen. Apples, berries, pomegranates, melon, avocado, nectarines. Vegetables like asparagus, kale, spinach, leeks, onions, spring onions, garlic, radishes, cabbage, ginger, salad and cooking potatoes are a fantastic source as their skin forms a gel structure that resists normal digestion in the stomach.

* * *

USEFUL ADDRESSES

Bowel Cancer UK
7 Rickett Street
London SW6 1RU
Tel: 08008403540
Website: www.bowelcanceruk.org.uk

Coeliac UK
Suites A–D, Octagon Court
High Wycombe
Bucks HP11 2HS
Helpline: 08704448804
Website: www.coeliac.org.uk

Colostomy Association
15 Station Road
Reading RG1 1LG
Helpline: 08003284257/08005876744
Website: www.colostomyassociation.org.uk

The Gut Trust (Formerly the IBS Network)
Unit 5, 53 Mowbray Street
Sheffield S3 8EN
Tel: 01142723253
Website: www.theguttrust.org

CORE (The Digestive Disorders Foundation)
3 Street Andrew's Place
London NW1 4LB
Website: www.corecharity.org.uk

Cancer Research UK
P.O Box 123
Lincoln's Inn Fields
London WC2A 3PX
Tel: 02072420240
Website: www.cancerresearchuk.org

British Nutrition Foundation
High Holbon House
52–54 High Holbon
London WC1V 6BQ
Tel: 02074046504
Website: www.britishnutrition.org.uk

NHS Stop Smoking/Alcohol
Helpline: 08001690169
Website: http://gosmokefree.nhs.uk

Crohn's & Colitis UK
1 Bishop's Square
Hatfield Business Park
Hatfield AL10 9NE
Tel: 01727830038
Website: www.crohnsandcolitis.org.uk

Cancerbackup
3 Bath Place
Rivington Street
London EC2A 3JR
Tel: 08088001234
Website: www.cancerbackup.org.uk

Allergy UK
3 White Oak Square
London Road
Swanley
Kent BR8 7AG
Tel: 01322619898
Website: www.allergyuk.org

Pancreatic Cancer UK
Westminster Tower
3 Albert Embankment
London SE1 7SP
Tel: 02035357090
Website: www.pancreaticcancer.org.uk

British Liver Trust
Venta Court
20 Jewry Street
Winchester
SO23 8FE
Tel: 08006527330
Website: www.britishlivertrust.org.uk

REFERENCES AND RESOURCES

Alcock. J, et al. *Is Eating behavior manipulated by the gastrointestinal microbiota?* 2014

Denby Nigel. *Nutrition for Dummies.* For Dummies, 2010

Dr Megan Arroll and Professor Christine Dancey. *Irritable Bowel Syndrome.* Hammersmith Health Books, 2018

Andrew Weil. *Eating Well for Optimum Health: The Essential Guide to Food, Diet and Nutrition.* Sphere, 2008

David. Frahm and Anne Frahm. *Healthy Habits.* Jeremy. P. Tarcher, 1998

Enders. Guilia. *GUT.* Scribe UK, 2017

Haas. M. Elson. *Staying Healthy with Nutrition.* Celestial Arts, 2016

Tim Taylor (2020, Oct 12) *Digestive System.* Retrieved from www. innerbody.com

Lin Chang (2017, Dec) *Digestive Illnesses.* Retrieved from www.ncbi. nlm.nih.gov

NHS (2020, Jan 22) *How our digestive system works and keeping it healthy.* Retrieved from www.gosh.nhs.uk

Better Health (2014, Aug 08) *Common problems in the digestive system.* Retrieved from www.betterhealth.vic.gov.au

John Hopkins (2018, Feb) *Eating for your gut.* Retrieved from www. hopkinsmedicine.org

INDEX

More Books by Josephine Spire

SELF-HYPNOSIS AND POSITIVE ASPIRATIONS
ISBN 978-1-84716-499-5 £9.99

Hypnosis is the gentle healer, no chemicals, no side effects, and it puts the patient in a state that holds great potential for healing by giving the patient access to the subconscious mind. Self- Hypnosis and Positive Affirmations is a book about how hypnosis combined with positive affirmations can be powerful in treating several physical, psychological, stress related disorders, phobias and promoting sporting performance among others.

This book is original and practical and will benefit anyone who wishes to investigate further. More and more people are beginning to realise and appreciate the healing power of hypnosis and affirmations

MIND POWER AND HEALTHY EATING

ISBN 978-1-84716-589-3   £9.99

The mind is a powerful tool in holding the power to the way we think, feel and act. Therefore, it's only fair to say that good health stems from the mind and this is where all the changes must begin. Mind power and Healthy Eating is a book about simple yet very powerful mind power techniques that are focused on promoting overall health whilst helping you to eat healthy, lose weight, maintain it as well as keeping active.

More and more people are giving up on diets than ever before because they have come to a conscious conclusion that diets don't work. In this book you will learn that when you access the power of the mind, you access well-being.

UNDERSTANDING AND MANAGING DEPRESSION AND STRESS

ISBN 978-1-84716-685-2 £9.99

Depression is a traumatic and cruel illness to suffer from, it can affect anyone regardless of their age, gender, race, or status. The list of famous and successful men and women who have suffered from depression is drawn out.

Depression is a common illness which affects 150 million people worldwide. Most people when depressed describe their feelings as despondent, almost like being under a constant dark cloud encompassed by a lot of sadness, irritability, frustration, negativity, and hopelessness among other symptoms, causing immense distress to the sufferer and their loved ones.

Throughout this book you will learn various mind techniques and coping skills that you can practice as part of your self-help to overcome depression and stress

STOP SMOKING NOW – THE SURVIVAL GUIDE
The Complete Guide to Quitting a Deadly Habit!

ISBN 978-1-84716-798-9   £9.99

An Emerald guide to

STOP SMOKING NOW
The Survival Guide

Josephine Spire

Studies show that few people understand the specific risks of tobacco use. Many smokers who are aware of the dangers of smoking want to quit the toxic habit but carry on anyway. The question is why do they carry on smoking? Is it because they are deeply addicted to the nicotine in the tobacco? Is the habit no longer controllable? Or is it the brainwashing that's sabotaging their ability to stop smoking?

The purpose of this book is to simply guide people to use the powers that they already possess to quit smoking and become non- smokers for good. The book is simple, short and to the point that the reader will find it easy to digest the information and how to use the diverse techniques provided for stop smoking.